£32-99
Ill
6/11

Programmable Logic Controllers and their Engineering Applications

Second Edition

To

4

Programmable Logic Controllers and their Engineering Applications

Second Edition

Alan J Crispin

Principal Lecturer
School of Engineering, Faculty of Information and Engineering Systems
Leeds Metropolitan University

The McGraw-Hill Companies

London · New York · St Louis · San Francisco · Auckland · Bogotá
Caracas · Lisbon · Madrid · Mexico · Milan · Montreal · New Delhi
Panama · Paris · San Juan · São Paulo · Singapore · Sydney · Tokyo · Toronto

Published by

McGraw-Hill Publishing Company

SHOPPENHANGERS ROAD, MAIDENHEAD, BERKSHIRE, SL6 2QL, ENGLAND

Telephone 01628 502500

Fax 01628 770224

British Library Cataloguing in Publication Data
Crispin, Alan J. (Alan John)
   Programmable logic controllers and their engineering
   applications. – 2nd ed.
   1. Logic circuits   2. Programmable logic devices
   3. Programmable controllers
   I. Title
   621.3'95

ISBN 0077093178

Library of Congress Cataloging-in-Publication Data
Crispin, Alan J.
   Programmable logic controllers and their engineering applications
/ Alan J. Crispin. – 2nd ed.
      p.   cm.
   ISBN 0–07–709317–8 (pbk. : alk. paper)
   1. Programmable controllers.   I. Title.
TJ223.P76C75 1997   96–27718
629.8'9–dc20   CIP

45 B&B 1

McGraw-Hill

A Division of The McGraw-Hill Companies

Printed and bound by Bell and Bain Ltd., Glasgow

Contents

Acknowledgements

I would like to extend my thanks to all the people who have in any way contributed to this book, in particular Bob Ward, Tarsen Hunjab, Boris Pokric, Maja Pokric, Jeff Cowey, Ibrani Lavdrus and Don Crispin. My special thanks go to Pam for her help in the preparation of the manuscript.

This book is dedicated to Danielle and Olivia.

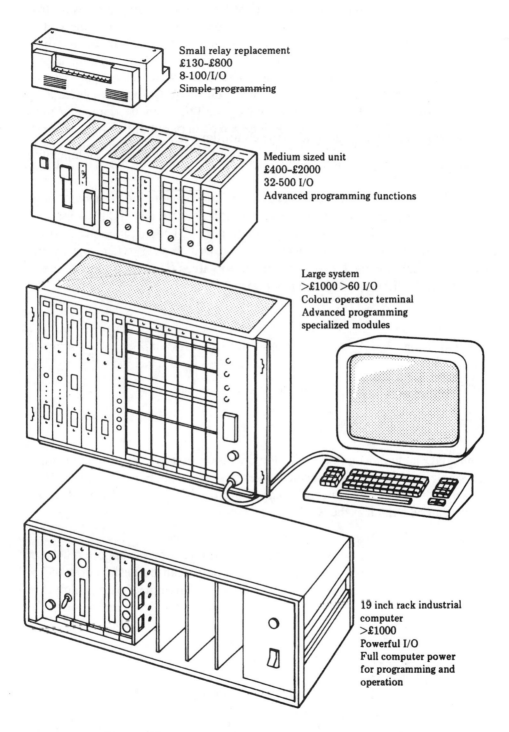

Small relay replacement
£130–£800
8-100/I/O
Simple programming

Medium sized unit
£400–£2000
32-500 I/O
Advanced programming functions

Large system
>£1000 >60 I/O
Colour operator terminal
Advanced programming
specialized modules

19 inch rack industrial
computer
>£1000
Powerful I/O
Full computer power
for programming and
operation

Figure 1.2 Types of PLC with approximate cost indicated. *Source*: Florence, D.H., 'PLCs aid low cost automation', *Professional Engineering*, July 1988, p. 32.

The field of logic is concerned with systems that work on a straightforward two-state basis. A common electric light switch can be either *on* or *off* and these alternate possibilities can be labelled as *true* and *false* or as 1 and 0 (binary form) respectively. A Boolean variable (i.e. a logical variable) such as A can be used to represent any switch-like element, which can have one of two states. For example, it is possible to define that A=1 when the switch *on* and A=0 when the switch is *off.*

Any condition in which there are two possibilities can be defined using a Boolean variable. For example, a workpiece can be in or out of position. This may be represented as a simple binary statement B=1 (workpiece is in position) or B=0 (workpiece is not in position). Another example is a lamp which can be either on or off.

If the Boolean variable A is used to represent the *on* and *off* positions of a single throw normally open (abbreviated to N/O) switch the variable NOT A represents a normally closed (abbreviated to N/C) switch. The NOT logic function inverts the state of the input. The variable NOT A can be written as $\bar{A}$, where the overbar denotes negation. Figure 1.3 tabulates the truth values (i.e. 0 and 1) of A against those of NOT A for the two switch positions.

Switch position	A	$\bar{A}$
Off	0	1
On	1	0

Truth table for the NOT function

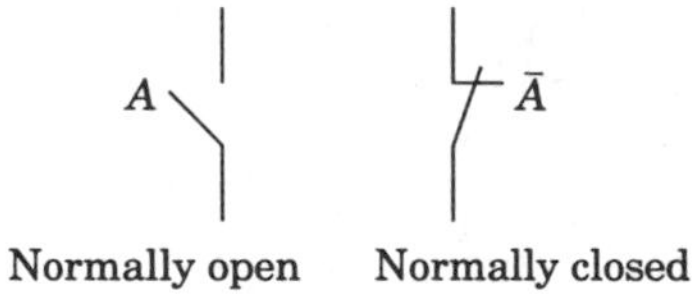

Normally open   Normally closed

Figure 1.3   The NOT function.

The AND logic function describes the operation of two normally open single-pole, single-throw switches connected in series as shown in Fig. 1.4. Current flows only when both the two switches A and B are on, i.e. when A=1 *and* B=1. The output *f* can be written as the Boolean expression:

$$f = AB$$

A	B	f = AB
0	0	0
0	1	0
1	0	0
1	1	1

Truth table for the AND function

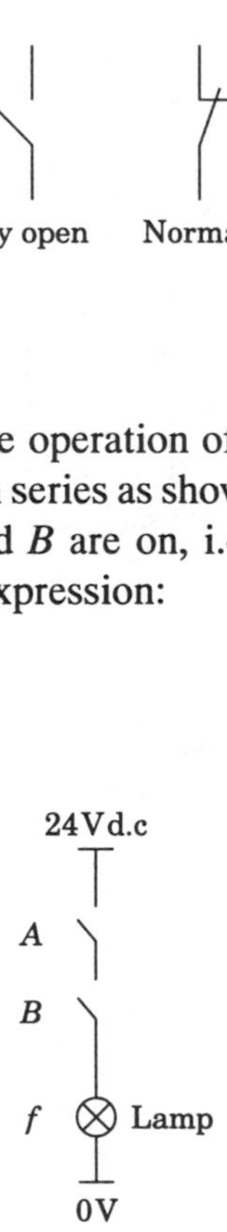

24Vd.c

Figure 1.4   The AND function.

Note that this is not a multiplication but the logical notation used to mean that f is 1 if A is 1 and B is 1. For an AND function with three inputs A,B,C (e.g. three N/O switches in series) the Boolean expression for the output f would be written as

$$f = ABC$$

The OR logic function describes the operation of two normally open switches connected in parallel as shown in Fig. 1.5. Current flows when either switch A or switch B is in the on position. The Boolean notation for A OR B is $A+B$.

A	B	$f = A + B$
0	0	0
0	1	1
1	0	1
1	1	1

Truth table for the OR function

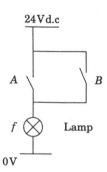

Figure 1.5   The OR function.

1.3   The laws of Boolean algebra

The above introduction has established the link between switching circuit diagrams and Boolean expressions. Rev. George Boole (1815–64) developed laws to analyse and construct logical statements. Boolean algebra deals with two-valued variables and is useful when analysing switching circuits such as ladder diagrams.

The laws of Boolean algebra consist of postulates and theorems. The four postulates which deal with the combination of one Boolean variable and with the constants 0 and 1 are

$$A + 0 = A \tag{1.1}$$
$$A + 1 = 1 \tag{1.2}$$
$$A0 = 0 \tag{1.3}$$
$$A1 = A \tag{1.4}$$

The two postulates that deal with one variable with itself are called the idempotent laws and are

$$A + A = A \tag{1.5}$$
$$AA = A \tag{1.6}$$

The commutative laws emphasize the fact that the position of the variables in an expression is not important. The commutive laws are

$$A + B = B + A \tag{1.7}$$
$$AB = BA \tag{1.8}$$

The associative laws deal with the use of brackets and can be expressed as

$$(A + B) + C = A + (B+C) \tag{1.9}$$
$$(AB)C = A(BC) \tag{1.10}$$

The distributive laws show how factors are combined and can be written as

$$(AB) + (AC) = A(B+C) \tag{1.11}$$
$$(A+B)(A+C) = A + (BC) \tag{1.12}$$

The complementarity and involution laws involve the NOT function. The complementarity laws are

$$A + \bar{A} = 1 \tag{1.13}$$
$$A\,\bar{A} = 0 \tag{1.14}$$

The involution law is

$$(\bar{\bar{A}}) = A \tag{1.15}$$

Two other postulates that involve the NOT function are called De Morgan's theorems which are

$$\overline{(AB)} = \bar{A} + \bar{B} \tag{1.16}$$

$$\overline{(A + B)} = \bar{A}\,\bar{B} \tag{1.17}$$

The function A AND B all negated (i.e. $\overline{(AB)}$) is called the NAND function. NAND is an abbreviation for NOT AND. Similarly, the function A OR B all negated (i.e. $\overline{(A + B)}$)) is called the NOR function. NOR is an abbreviation for NOT OR.

There are two branches of logic called combinational logic and sequential logic. A combinational logic system is one in which the output is a direct and unique consequence of the input conditions. A sequential logic system is one that depends on the sequence in which the inputs occur. This introduction has dealt with combinational logic. Sequential logic devices (e.g. flip-flops, shift registers) are discussed in Chapter 6.

1.4 Ladder diagrams

PLCs were developed to offer a flexible (e.g. programmable) alternative to conventional electrical circuit relay-based control systems built using discrete devices. The terminology and other concepts used to describe the operation of a PLC are based on conventional relay control terminology. The relationship is such that inputs are referred to as contacts, outputs are referred to as coils and memory elements (bits) are referred to as auxiliary relays.

The International Electrotechnical Commission (IEC) advocates five programming methods for PLCs which are fully discussed in Chapter 5. Of these five, the predominant programming method used by all mainstream PLC systems is the ladder diagram method. This is a graphical programming technique which has been evolved from conventional electrical circuit relay logic control methods. PLC systems provide a design environment (either in the form of software tools running on a host computer terminal or a hand-held LCD graphic programming console) which allow ladder diagrams to be developed and diagnosed.

In its simplest form, a ladder diagram is a network of contacts and coils bounded on the left and (optionally) on the right by vertical lines called power rails. The key features of a ladder diagram are:

1. Contacts represent the states of Boolean variables. For example, a contact called 'start_fan' might represent the ON and OFF state of a switch which is used to initiate a ladder rung which operates a fan. The ladder symbols for normally open and normally closed contacts are $||$ and $|\!/|$ respectively.
2. A typical ladder rung is a horizontal line of contacts which start from the left power rail and which provide the logic to operate a coil (or coils). A coil is a Boolean variable on the right of a ladder rung which can be set to a true state when the contacts connecting it to the left power rail are on. The ladder symbol for a coil is a pair of parenthesis ( ).
3. When contacts of a ladder rung are in the true state notional power is deemed to flow from the left power rail through the contacts to operate a coil (or coils) at the right-hand end of the rung. The drawing of the right power rail is optional as its use is implied.
4. The normal convention for evaluating ladder rungs is from top to bottom. Each ladder rung is scanned and evaluated one after the other starting from the top. The cyclic scan based operation is further discussed in Chapter 2.
5. Function block elements such as timers, counters and shift registers can be connected into a ladder diagram provided that their inputs and outputs are Boolean variables. Function block elements are discussed further in Chapters 5 and 6.

An example ladder diagram is shown in Fig. 1.6. A ladder diagram is drawn as a set of rungs with each one representing a control action. An input contact represents the state of a Boolean variable. In the first ladder rung the inputs labelled A and B are connected in series and represent a two-input AND function. The output f_1 is 1 (true) if A is 1 and B is 1. In the notation of Boolean algebra the AND function is written as $f_1 = AB$.

The second ladder rung represents a three-input (i.e. the inputs C, D and E) AND function in which two normally closed (N/C) contacts are used. The third ladder rung represents a two-input OR function. This time the contacts have been labelled (identified) as switch 1 and switch 2 and the output as the Boolean variable 'pump_motor'.

The names that are used to reference Boolean variables are called 'identifiers'.

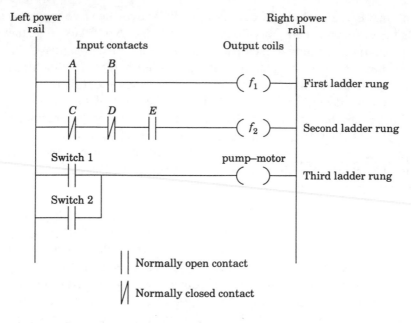

Figure 1.6   An example ladder diagram.

In Fig. 1.6 the names switch 1, switch 2 and pump_motor which refer to input and output devices are examples of identifiers. The naming of contacts and coils in this way aids the readability of a ladder diagram. However, note that each identifer has to be assigned a unique memory address used to identify an I/O point. The binary, octal and hexadecimal number systems (see Appendix 2) are widely used in the direct referencing of I/O points.

References

1. *Programmable Controllers—Part 1: General Information*, International Electrotechnical Commission, IEC 1131-1, 1992 (also British Standard BS EN 61131:1994).
2. D.H. Florence, 'PLCs aid low cost automation', *Professional Engineering*, July 1988, p. 32.

2
Design, structure and operation

A modern PLC is a microprocessor-based control system designed to operate in an industrial environment. PLCs are programmed to sense, activate and control industrial equipment and therefore incorporate a large number of I/O points which allow a wide range of electrical signals to be interfaced. In this chapter some of the important concepts behind the design of PLC hardware are presented.

2.1 PLC architecture

The internal hardware and software configuration of a PLC is referred to as its architecture. Being a microprocessor-based system the design of a PLC is based around the following building blocks:

- Central processing unit (CPU)
- Memory
- Input/output interface devices

These elements are semiconductor integrated circuits (ICs). They are inter-connected by means of a bus as shown in Fig. 2.1. A bus is a group of lines over which digital information (e.g. 8 bits of data) can be transferred in parallel. In most systems there are three distinct buses: the data bus, the address bus and the control bus.

The operation of a PLC system is determined by a program written by the user. High-level programs such as ladder diagrams are converted into binary number instruction codes so that they can be stored using digital memory devices such as RAM (random access memory) or programmable ROM (read only memory). Each successive instruction is fetched, decoded and executed by the CPU.

2.2 CPU

A CPU generally takes the form of a single microprocessor device. The function of the CPU is to control the operation of memory and I/O devices in the system and to process data in accordance with the program. To do this it requires a clock signal to sequence its internal operations.

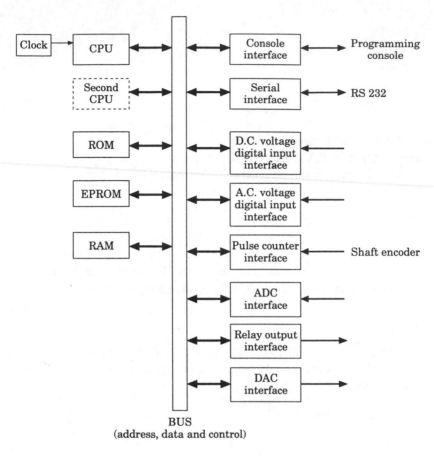

Figure 2.1   Simplified architecture of a PLC system.

The essential elements of a microprocessor are:

● Registers
● Arithmetic logic unit (ALU)
● Control unit

A register is a byte (8 bits), word (16 bits) or long word (32 bits) of memory which is part of the microprocessor as opposed to general purpose memory. A register is used for temporary storage of data and addresses within the CPU. The ALU performs arithmetic and logical operations such as addition and subtraction on data stored in registers. The control unit is basically a set of counters and logic gates which is driven by the clock. Its function is to control the units within the microprocessor to ensure that operations are carried out in the correct order.

Some CPU registers are accessible to the programmer while others are not. Some registers common to most microprocessor devices are:

- Data registers
- Address registers
- A program counter
- A flag register
- A stack pointer

Data registers hold data which is to be operated on by the ALU. A bit pattern moved into a data register can be added, subtracted, compared, etc., with another bit pattern stored in a separate data register.

Each memory location in RAM and ROM for storing data is given an unique address. CPU address registers can be used by the programmer to specify source and destination addresses of data items to be manipulated. The program counter (also called the instruction pointer) holds the address of the next instruction to be executed and is automatically incremented by the CPU.

A flag register is a collection of single-bit status indicators (flags) that holds information about the result of the most recent instruction that affects them. Commonly used flags are:

- *Carry bit*   Set to 1 if a binary addition operation produces a carry or a subtraction operation produces a borrow.
- *Zero bit*   Set to 1 whenever the result of an operation is zero.
- *Negative bit*   In signed binary arithmetic it indicates the sign (positive or negative) of the result.
- *Overflow*   Set to 1 when the result of an arithmetic operation cannot be represented in the specified register size.

A stack is a variable length data structure in which the last data item inserted is the first to be removed. A stack pointer register contains the address of the top element of the stack. The CPU uses the stack to store subroutine return addresses for example.

The execution time of each instruction code takes a specific number of clock cycles. The clock cycle time is the reciprocal of the clock frequency. For example, a 10 MHz clock has a clock cycle of 0.1 μs.

2.3   Memory

Memory devices store groups of binary digits (usually bytes) at individual locations identified by their own unique addresses. Memory devices are ICs having an address input (commonly 16 bits wide) and an in/out data port (commonly 8 bits wide). There are two main types of memory, namely RAM (random access memory) and ROM (read only memory). The storage capacity of a memory device is determined by the number of binary digits it can hold. A 1K byte memory device is capable of storing 1024 (i.e. 2^{10}) bytes.

A large proportion of the total addressable memory space is devoted to RAM, which is capable of having data written to it and read from it by the CPU. RAM is used for program and data storage. A backup battery supply is needed to retain the memory contents of RAM as stored data is lost when the power is removed.

There are two main types of RAM, namely SRAM (an abbreviation for static RAM) and DRAM (an abbreviation for dynamic RAM). Static RAM holds data in flip-flop type cells which stay in one state until rewritten. SRAM has a fast data access time of typically 10–20 ns. Dynamic RAM requires special circuitry to provide a periodic refresh signal in order to maintain the stored data. This is because DRAM technology stores data in capacitor type cells which must be periodically refreshed as capacitors discharge as time increases. DRAM is cheaper to manufacture but because a refresh signal is used access time is slower than that of SRAM, typically 50–60 ns.

The PLC operating system (i.e. the program that allows the user to develop and run applications software) is stored in a type of memory referred to as ROM. Once programmed, this type of memory can only be read from and not written to but does not lose its contents when the power is removed.

Common types of read only memory which are user programmable are EPROM (erasable programmable read only memory) and EEPROM (electrically erasable programmable read only memory). EPROMs are programmed using a dedicated programmer and erased by exposing a transparent quartz window found in the top of each device to ultraviolet light. The erasing process clears all memory locations and takes between 10 and 20 minutes. EEPROM is erased using electrical pulses rather than ultraviolet light. Most PLC systems provide facilities that allow application software to be stored in either EPROM or EEPROM.

A memory map is used to indicate how address locations are allocated to ROM, RAM and input/output devices. If I/O devices are placed in the memory address

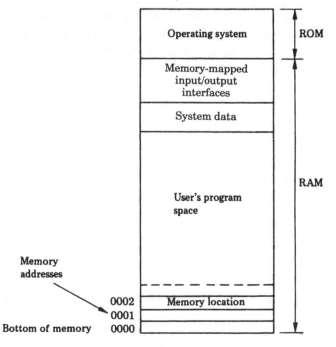

Figure 2.2 Memory map.

space they are said to be memory mapped and access to I/O points is via load and store memory instructions. This has the advantage that a common set of instructions are used for memory and I/O operations. A memory map of a typical PLC system is shown in Fig. 2.2.

2.4   Bus

A bus can be considered as a set of lines over which digital information (e.g. 16-bit address or 8-bit data) can be transferred in parallel. In most systems there are three distinct buses:

● Data bus
● Address bus
● Control bus

The data bus is a bidirectional path on which data can flow between the microprocessor, memory and I/O. An 8-bit microprocessor has a data bus which is 8 lines wide. A 16-bit microprocessor has a data bus which is 16 lines wide.

The address bus is a unidirectional set of lines which carry binary number addresses. Addresses are generated by the CPU during the execution of a program to specify the source and destination points of the various data items to be moved along the data bus. An address identifies a particular memory location or I/O point.

The control bus consists of a set of signals generated by the CPU to control the devices in the system. An example is the read/write control line which selects one of two operations, either a write operation where the CPU is outputting data on the data bus or a read operation where the CPU is inputting data from the data bus.

Digital devices sharing a bus must be tri-state. This means that when the output lines are not in use they are put into a high impedance state so that they will not load the bus. Therefore an output line can have one of three possible conditions, which are logic 0 (low), logic 1 (high) and output disconnected. Tri-state devices incorporate a chip select (also called chip enable) input which is used to isolate its data output lines from the bus. When the chip select line is not active the data output lines are placed in to the high impedance or tri-state. The control bus co-ordinates the connection of the various devices to the data bus.

2.5   Input/output interfaces

Microprocessors are supported with special purpose peripheral I/O ICs to enable them to interface with external devices such as keypad displays, sensors and load actuators. Examples of special purpose I/O ICs include keyboard controllers, programmable parallel interface devices, programmable serial interface devices and counter/timer devices.

As far as the user is concerned, it is the front end circuits to which sensors and actuators are connected that is important. Input points can include the following types of interface:

- D.C. voltage digital input circuit
- A.C. voltage digital input circuit
- Pulse counter circuit
- ADC interface

Output points can include the following types of interface:

- Relay output circuit
- Transistor output circuit
- Triac output circuit
- DAC interface

2.5.1   D.C. VOLTAGE DIGITAL INPUT CIRCUIT

Figure 2.3 illustrates typical 24 V d.c. input circuits for connecting current sinking and sourcing input devices. With the sink input interface the input device when turned on connects the circuit to the 0 V line of the d.c. supply. Current then flows through the status LED (light-emitting diode) used to indicate the current logic state of the input point and the opto-isolator.

An opto-isolator combines an LED and photoelectric transistor. When current is passed through the LED it emits light, causing the photoelectric transistor to turn on. Provided a separate supply is used for the LED and photoelectric transistor circuits a very large degree of isolation is maintained between the two components.

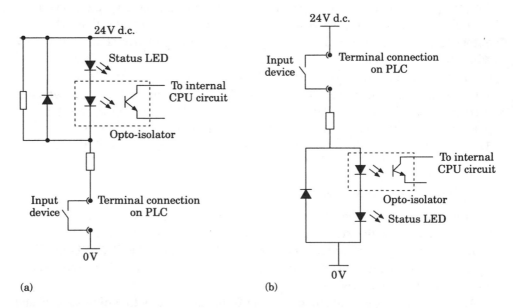

(a)

(b)

Figure 2.3   D.C. voltage digital input circuits: (a) sink input circuit and (b) source input circuit.

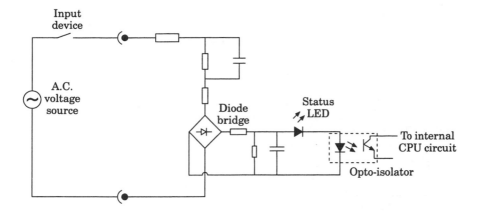

Figure 2.4   A.C. voltage digital input circuit.

With the current sourcing interface the input device, when turned on, connects the circuit to the positive polarity of the supply (i.e. 24 V). Current flows from the supply through the status LED and opto-isolator circuit.

2.5.2   A.C. VOLTAGE DIGITAL INPUT CIRCUIT

Figure 2.4 illustrates an a.c. voltage digital input circuit. A full-wave diode bridge circuit is used to convert the a.c. input signal into a rectified d.c. signal. The status LED and opto-isolator are turned on by the rectified d.c. signal.

2.5.3   PULSE COUNTER INTERFACE

Special purpose ICs are available which integrate a high-speed digital pulse counter circuit and buffer memory. Input modules based around these ICs can be used to read pulse streams from an input device such as a shaft encoder. The advantage of using such modules is that high-speed pulse counting is not slowed down by the cyclic scan operation of a PLC because the counter circuit operates independently. Provided the memory buffer is accessible to the ladder program it can be read at the point in time when the pulse count value is required. It is also possible to use a technique whereby the high-speed counter circuit interrupts the CPU when the count value reaches some pre-defined limit.

2.5.4   ANALOGUE TO DIGITAL CONVERTER (ADC) INTERFACE

An analogue input circuit will incorporate an analogue to digital converter (ADC) device, as shown in Fig. 2.5. An ADC accepts an analogue input signal and converts it into an output binary value that corresponds to the level of the analogue input. Most ADC devices incorporate a start convert (SC) pin, an end of convert (EOC) pin and a sample and hold circuit. When the start convert signal is pulsed the ADC samples and holds (stores) the analogue value input to the device at that

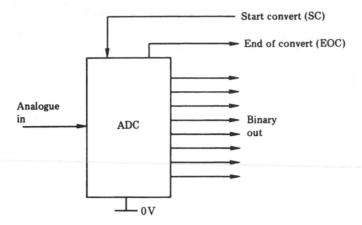

Figure 2.5   Analogue to digital converter (ADC).

time. The ADC converts the analogue signal into a digital value and then produces an end of convert signal when the conversion is complete.

The time taken by the ADC to convert an analogue signal into a digital number is called the conversion time. There are a number of different types of ADC devices commercially available which work on different principles. One common type uses a technique called successive approximation to obtain short conversion times. This involves comparing the output of a digital to analogue converter (DAC) with the voltage to be converted by making a series of successive guesses (i.e. approximations) at the value of the binary number required. The fastest type of ADC is the flash converter which determines simultaneously all the bits for the digital number representing the analogue input level.

According to sampling theory, an ADC should sample an input signal at least twice as fast as the input's highest frequency component. If the input signal is sampled too slowly, aliasing occurs. In this case, a high frequency signal is represented by an erroneous lower frequency value. A technique to prevent an aliasing error condition occurring is to precede the ADC with an anti-aliasing filter which band limits the input signal to half the sample frequency.

Speed of operation is not the only consideration when selecting an ADC. An 8-bit ADC approximates the analogue input voltage into one of 2^8 or 256 discrete levels called quantization levels. If there are 256 quantization levels including zero, there will be 255 steps between them. This means that, for an 8-bit ADC designed so that the maximum of the analogue input is limited to 5 V, the quantization interval (i.e. resolution) is 5/255 or 19.6 mV. The only way to obtain more quantization levels is to increase the number of bits used in the conversion. For example, a 12-bit ADC will have 2^{12} or 4096 levels.

The input range of an analogue voltage interface can be unipolar (0–10 V for example) or bipolar (± 10 V for example). In addition, input channels can often be configured as either single-ended or differential inputs. A single-ended input has one terminal connected to 0 V so that the signal varies with respect to 0 V. A differential input measures the difference between two signal leads. Differential inputs can provide noise immunity as a noisy signal occurring equally on both signal leads is cancelled out when the voltage difference is measured.

In industrial applications analogue currents rather than voltages are often used. This is because a voltage applied to one end of a cable is reduced by the resistance of the cable whereas the current remains fixed. The resistance of a cable is proportional to its length and so the reduction of voltage increases with length. A commonly used current input range is 4–20 mA. By using this range a broken cable gives a result of 0 mA, which is identified as an error. The input of current to an ADC is performed by measuring the voltage across a known resistance through which the current is passing and applying Ohm's law.

2.5.5   RELAY OUTPUT CIRCUIT

Figure 2.6 illustrates the circuit for a relay output interface. The NPN transistor is used to switch current through the relay coil to close its contact. The transistor is controlled by the internal circuit of the PLC. The diode is connected across the relay coil to protect the transistor from the effects of back e.m.f. This is the reverse voltage developed in the relay coil which causes an inductive current which opposes the normal flow of current. Both a.c. and d.c. supplied loads can be connected through the relay output terminals.

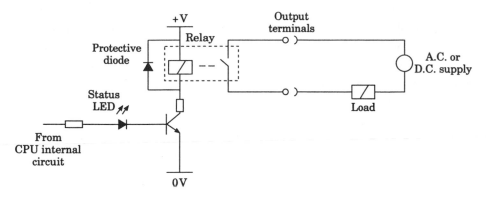

Figure 2.6   Relay output circuit.

2.5.6   TRANSISTOR OUTPUT CIRCUIT

A transistor output circuit is used for switching d.c. voltages. Figure 2.7 shows a typical opto-isolated NPN transistor switch output circuit. The principle of operation of a transistor switch ideally relies on there being:

● An open circuit (i.e. infinite resistance) between the collector and emitter when the base–emitter circuit is not forward biased

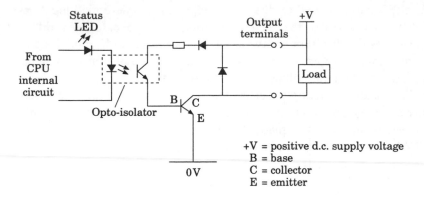

Figure 2.7   NPN transistor switch circuit.

● A short circuit (i.e. zero resistance) between the collector and emitter when the base–emitter circuit is forward biased

The ideal switching action of a transistor requires that the base current be large enough for the collector current to reach its maximum or saturation value. At saturation, the voltage drop between the collector and emitter is very small. Consequently, the collector is close to 0 V and the collector current may be roughly determined from the load resistance and supply voltage.

2.5.7   TRIAC OUTPUT INTERFACE

A triac output device is used for switching a.c. voltages. Triac devices that incorporate a zero crossing circuit to monitor the a.c. cycle have the advantage that they turn off at zero current. Inductive loads should be turned off at a zero current crossing point to prevent interference. A typical opto-isolator-based triac output circuit is shown in Fig. 2.8.

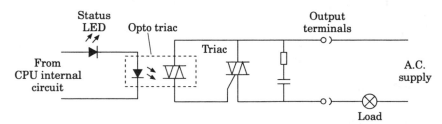

Figure 2.8   Triac output circuit.

2.5.8   DIGITAL TO ANALOGUE CONVERTER (DAC) INTERFACE

An analogue output module will incorporate a digital to analogue (DAC) device as shown in Fig. 2.9. A DAC converts a binary number input into a proportional

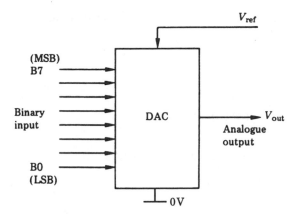

Figure 2.9   Digital to analogue converter (DAC).

analogue level. The analogue output is produced from a reference voltage V_{ref}. The binary number input to the DAC determines what fraction of V_{ref} is presented at the output. For the 8-bit DAC shown in Fig. 2.9 the output is calculated from the equation

$$V_{out} = V_{ref}(b_7/2 + b_6/4 + b_5/8 + b_4/16 + b_3/32 + b_2/64 + b_1/128 + b_0/256)$$

where the bits b_0 to b_7 can take the values 0 or 1 and are the binary inputs. Here b_7 is the most significant bit (MSB) and b_0 the least significant bit (LSB).

An 8-bit DAC has 2^8 or 256 quantization levels and 255 steps between them. This means that for an 8-bit DAC with a reference voltage of 10 V the quantization interval (i.e. resolution) is 10/255 or 39.2 mV. A 12-bit DAC with a reference voltage of 10 V has a quantization interval of 10/4095 or 2.4 mV.

The output of a DAC begins to change when it receives a new data value at the input. The period required before the output is valid is called the settling time. Generally, settling times are small, being specified in microseconds.

2.6   Input/output assignment

Each input and output connection point on a PLC has an address used to identify the I/O bit. Each manufacturer uses a proprietary identification system which is related to the type and number of input/output options possible.

The IEC 1131-3 programming languages standard suggests a method for the direct representation of data associated with the inputs, outputs and memory of a programmable logic controller.[1] This is based on the fact that a PLC memory is organized into three regions. These are input image memory (I), output image memory (Q) and internal memory (M). Any memory location including those representing I/O bits can be referenced directly using

% (first letter code) (second letter code) (numeric field)

where the % character indicates a directly referenced variable. The first letter code specifies the memory region:

I = input memory
Q = output memory
M = internal memory

The second letter code specifies how memory is organized:

X = bit
B = byte
W = word
D = double word
L = long word

If a second letter code is not given it is assumed to be a bit.

The numeric field component is used to identify the memory location. It supports the concept of I/O channels since numeric fields can be separated using a period.

Some examples of direct referencing of I/O memory as advocated in the IEC 1131-3 standard are:

%X1 (* input memory bit 1 *)
%I1 (* also input memory bit 1 *)
%IB2 (* input memory byte 2 *)
%IX10.4 (* input byte address 10 bit 4 *)
%QX1 (* output memory bit 1 *)

Using names that identify the purpose of each contact and coil in a ladder diagram aids readability. For example, the contact given the name 'start_fan' clearly identifies the purpose of the input signal. This variable may be the input point directly referenced as %IX1 (e.g. input memory bit 1). Input and output memory locations can be assigned directly on a ladder diagram as shown in Fig. 2.10.

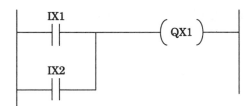

Figure 2.10   Assignment of I/O points.

2.7   Keyboard and display

User programs can be entered into RAM using a program console unit consisting of a keyboard and display. Some typical hand-held PLC programming consoles[2,3] incorporating liquid crystal displays for viewing ladder code and diagrams are shown in Fig. 2.11.

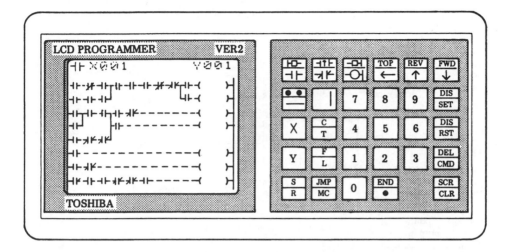

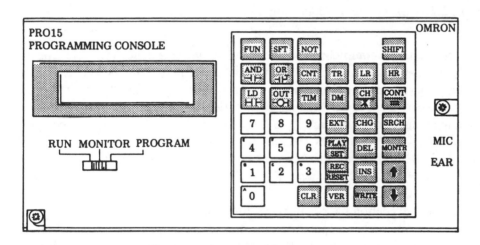

Figure 2.11 Typical PLC program consoles.

2.8 Program execution

The most common approach taken to executing a PLC program is to use a cyclic scan or main program loop such that periodic checks are made on the input values. Figure 2.12 illustrates the way a ladder program can be executed. The program loop starts by scanning the inputs to the system and storing their states in fixed memory locations referred to as input image memory. The ladder program is then executed rung by rung, starting at the first rung. The output states are determined by scanning the program and solving the logic of the various ladder rungs. The updated output states are stored in fixed memory locations referred to as output image memory. The output values held in memory are then used to set and reset the physical outputs of the PLC simultaneously at the end of the program scan.

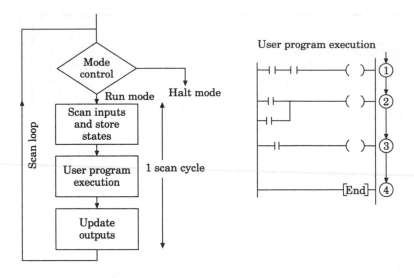

Figure 2.12   Executing a ladder program.

The time taken to complete one cycle is called the cycle or scan time. It depends on the length and complexity (e.g. the number and types of functions used) of the program. Consequently, manufacturers specify an average value. For example, the average execution speed for Mitsubishi F1 series PLCs is specified as $12\ \mu s$/step, with a maximum program capacity of 1000 steps.[4] This yields an average cycle time of 12 ms for 1000 steps. The average execution speed for Mitsubishi F2 series PLCs is specified as $7\ \mu s$/step, with a maximum program capacity of 2000 steps (F2-40,60M) and 1000 steps (F2-20M).[5] This yields an average cycle time of 7 ms for 1000 steps.

Scan-based execution has a number of limitations when the PLC system has to respond to events within a specified time period. PLC systems which accommodate interrupts can be used to spontaneously respond to a specific event such as an alarm. An interrupt is a special control signal to the CPU which tells it to stop executing the program in hand and start executing another program stored elsewhere in memory (i.e. interrupts the sequential execution of the ladder program).

2.9 Multitasking and multiprocessing

Advanced PLC systems incorporate a processor scheduler to enable multitasking and multiprocessing. Multitasking is the running of two or more tasks (also called processes) on a single processor such that they share processor time so that they appear to run in parallel. Different tasks of a program can be executed at different rates. Consequently, time critical tasks (e.g. the monitoring of the state of a limit switch) can be given a high priority and scheduled to be executed within a fixed time period.

Multiprocessing is the running of tasks or processes simultaneously on different processors. In this context a task or process is considered to be a separate code fragment which performs a discrete activity (a program organizational unit, or POU in IEC terminology). With a multiprocessing system, tasks such as 'closed-loop PID (proportional, integral, derivative) control' and 'ladder circuit control' can be mapped on to separate processors and run concurrently, communicating with one another.

2.10 Development systems

A PLC development system would normally comprise a host computer connected via a serial communications RS232C port to a target PLC. The host computer provides a software environment to perform editing, file storage, printing and program operation monitoring. Typically, the process of writing a program to run on the PLC consists of:

- Using an editor to write/draw a source program
- Converting the source program to binary object code which will run on the PLC's microprocessor
- Down loading the object code from the host PC to the PLC system via the serial communications port

The editor enables programs to be created and modified on the host computer in either graphical form, such as a ladder diagram, or text form, such as mnemonic code. Features such as cut and paste, copy program block and address search are standard.

A debugger is a software tool associated with an editor which allows the user to examine instructions and data to help get the bugs out of a program as it runs. A PLC system which is in active control of a machine or process is referred to as being on-line. With many systems it is possible to monitor any data bit while the PLC is on-line to check for correct operation.

Further information relating to characteristics relevant to the selection and application of programmable logic controllers, including information to be provided by the manufacturer, is covered in the International Standard (IEC 1131-2).[6]

References

1. *Programmable Controllers—Part 3: Programming Languages*, International Electrotechnical Commission, IEC 1131-3, 1993 (also British Standard BS EN 61131-3: 1993), p. 35.
2. *Programming Console*, CK-series Operation Manual, Omron, 1988.
3. *Hand-held Graphic Programmer*, HP911 Operation Manual, Toshiba, 1995.
4. *Melsec F1 Series Instruction Manual*, Mitsubishi Electric Corporation, 1986.
5. *Enhanced F2 Series Instruction Manual*, Mitsubishi Electric Corporation, 1986.
6. *Programmable Controllers—Part 2: Equipment Requirements and Tests*, International Electrotechnical Commission, IEC 1131-2, 1994 (also British Standard BS EN 61131-2: 1994).

3
Power and control circuits

3.1 Standardization of circuit diagrams

This chapter introduces some basic ideas relating to electrical control equipment used in conjunction with PLCs. Circuit diagrams are drawn to help the understanding of the operation of equipment. The IEC has defined a set of rules, graphical symbols and a referencing method which should be used when designing circuit diagrams and marking electrical control equipment.

A general point to make is that power and control circuits have been clearly labelled. Power circuits involve mains supply lines. Control circuits comprise components intended for use in low voltage installations. Programmable logic controllers are considered as components of a control system.

Standard graphical symbols are used to denote electrical and mechanical components. Each component symbol within a circuit diagram is marked with a unique reference label which directly connects it to a physical component. References have an alphanumeric and a numeric part with letters coming before the numbers. Some instances of alphanumeric references are F (protection device such as a fuse), M (motor), K or KM (contactor) and S (switch device for control circuits). Some examples of reference labels are:

- KM1 references contactor device number 1
- S2 references switch device number 2

The IEC graphic symbols for some principal elements and a list of IEC reference letters are shown in Fig. 3.1.

3.2 Circuit supply basics

Mains supply connections are labelled on many of the following circuit diagrams. Readers are warned that precautions must be taken to ensure safety when designing and constructing automation equipment with mains supply voltage. If due care and attention is not taken the results can be lethal. The aim of this section is to provide a short overview of power supply basics and terminal labelling on circuit diagrams.

24

Description of symbols	IEC symbols	
	Power	Control
Normally contact N/O	d	
Normally contact N/C	b	
Fuses		
Relays	Thermal	Magnetic
Operating coils	A1 A2	
Motors	M 3~	
Disconnect switches, isolators		
Circuit breakers		

Reference letters	
A	Assemblies and subassemblies (standard)
B	Transducers of a non-electric quantity into an electric quantity (or vice versa)
C	Capacitors
D	Timers, storage devices
E	Miscellaneous equipment
F	Protection devices (e.g. fuse)
G	Generators (for power supply)
H	Signalling devices
K	Relays and contactors
KA	Auxiliary contactors, control relays
KM	Main contactors
L	Inductances
M	Motors
N	Subassemblies (non-standard)
P	Meters, test devices
Q	Mechanical connecting devices for power circuits
R	Resistors
S	Mechanical connecting devices for control circuits
T	Transformers
U	Converters
V	Electronic valves, semi-conductors
W	Transmission lines, aerials
X	Terminals, plugs, sockets
Y	Electrically actuated mechanical devices
Z	Corrective loads

Figure 3.1 IEC graphical symbols and reference letters.

A three-phase supply is one in which alternating current (a.c.) is provided along three separate lines with 120° between the voltage phases of any two lines. Figure 3.2 illustrates the secondary of a three-phase transformer whose primary is supplied by the grid. The three phases are labelled as red (R), yellow (Y) and blue (B) or L1, L2 and L3 respectively. A fourth connection called the neutral (N) is made to the star point of the transformer and acts as a common return line for the currents in the three phases. In the United Kingdom the standard three-phase supply is 415 V at 50 Hz. This means that there are 415 V r.m.s. between any two phases such as L1 and L3. There are 240 V r.m.s. between a phase such as L3 and the neutral.

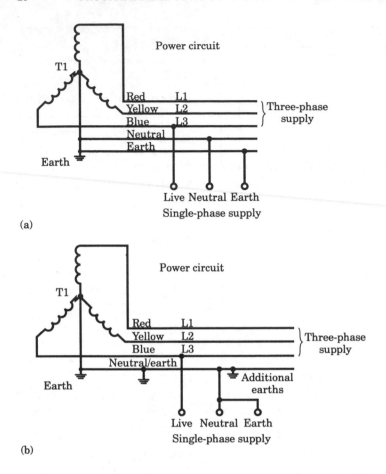

Figure 3.2   Mains supply basics: (a) TN–S system and (b) TN–C–S system.

The neutral point of the grid transformer is earthed and the distribution system is described as having an earthed neutral. The grid authority can make a connection to earth by burying a metal plate into the ground (i.e. earth) and connecting to this. The earth itself is taken as being a zero voltage point. The earth connection represents the electrical safety conductor.

A single-phase consumer supply consisting of live (L), neutral (N) and earth (E) lines is derived from the three-phase supply as shown in Fig. 3.2. The live line is connected to one of the three phases. The main earth terminal may be connected to the transformer earth by using a separate conductor as shown in Fig. 3.2(a). This is referred to as a TN–S system. Earthing systems are designated in the IEE regulations using the letters:

T = terre (French for earth)
N = neutral
C = combined
S = separate

Alternatively, the earth and neutral lines may be connected together at the grid authority's terminals so that the transformer's earth is connected to the consumer supply cable via the neutral conductor. In this case, the neutral line is earthed at a number of other points to reduce the risk of a break in the neutral line resulting in the loss of the earth return path to the three-phase grid transformer. The use of a number of separate earth points is referred to as protective multiple earthing (PME). This system is referred to as the TN–C–S system and is illustrated in Fig. 3.2(b).

With a single-phase supply the load is connected between the live and neutral conductors. The live conductor is 240 V r.m.s. with respect to earth. The neutral conductor will normally be a few volts above earth potential. This is due to the fact that load currents flow through the finite resistance of the neutral cable. Both the live and neutral conductors are classed as phase conductors since crossed connections (i.e. a fault in the installation wiring) can result in the neutral line being 240 V. The neutral conductor and live conductor should be treated with equal care.

The earth or protective conductor provides an alternative connection to the neutral of the supply and can therefore be used to provide protection against electric shock. For example, all accessible conductive parts of a piece of equipment such as the enclosure should be connected to the protective earth conductor of the supply. This provides a path for fault current should a fault cause a conductive part to become live.

To obtain a direct current (d.c.) supply from a single-phase mains connection requires a transformer and rectifier circuit as shown in Fig. 3.3. The d.c. power supply leads are often labelled + and − on circuit diagrams, which is supposed to relate to the positive and negative terminals marked on a battery. The negative side can be taken as common to a circuit and labelled 0 V. Consequently, a point labelled 0 V represents the d.c. return path.

The use of + and − marking can be misleading when labelling a split d.c. power supply (also called a dual or centre tapped d.c. supply) which has both a positive and a negative voltage supply with reference to 0 V. In this book, the positive and negative rails of a split supply are labelled +V and −V to indicate

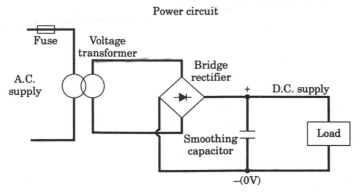

Figure 3.3   D.C. supply circuit.

the appropriate polarity relative to the common 0 V line. For example, a split supply marked +9 V and –9 V implies +9 V with reference to 0 V and –9 V with reference to 0 V.

3.3 Protection devices

The power circuit of an automatic control system must perform the functions of isolation, safety control, functional control and electrical protection which detects overloads and short circuits. An isolator provides isolation at the point of entry of the incoming supply and enables the supply to be disconnected so that work can be carried out on the load side of the circuit without danger. It can also incorporate fuses for short circuit protection. Safety control requires that all live conductors must be broken where a danger to persons exists, for example, in cases such as an emergency stop and supply failure. Functional control is the switching of the current on and off when required.

Protection devices must be used in electrical control circuits to detect any electrical and mechanical problems which can occur in the load. An example of a mechanical problem would be the stalling of a motor which results in a rapid increase in motor current. If the increase in motor current persists it would cause problems such as the heating of the motor windings and the melting of cable insulation materials with the risk of fire. In this case, a current overload protection device would inhibit motor operation and prevent these problems from occurring. However, with the motor there is an added complication in that a large starting current is initially required. Any protection device must allow the motor to start but also protect the motor by preventing its operation when an overcurrent occurs for too long a time period.

There are various types of protection device which are selected according to the type of protection required. A summary of the type of protection offered by various devices is shown in the Table 3.1.

Table 3.1 Protection devices

Type of protection required	Device
Short circuits	Fuse, circuit breaker
Small and sustained overcurrent protection	Thermal overload relays
Large overload protection	Electromagnetic overcurrent relays
Phase imbalance, single phasing	Differential thermal relays
Supply failure	Contactor with maintaining (auxiliary) contact
Overcurrent and undercurrent	Current measuring relays
Overlong motor starting	Thermal time delay relays

Figure 3.4 shows the graphic symbols of various types of multiple function protection devices used in the installation of automatic control equipment.[1] Figure 3.4(a) combines a fused isolator, thermal overload relay and contactor. Figure 3.4(b) illustrates the symbol for a motor circuit breaker protection device. Figure 3.4(c) illustrates the symbol for a manual starter with thermal and magnetic trips.

Power circuits

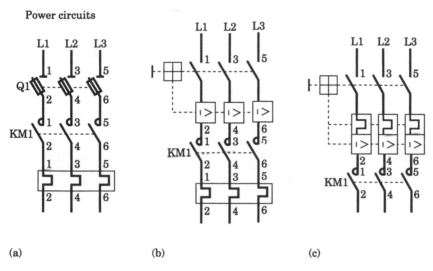

(a)   (b)   (c)

Figure 3.4 Graphical symbols for multiple function protection devices: (a) fused isolator, contactor and thermal overload relay, (b) motor circuit breaker, contactor and thermal overload relay and (c) manual starter with thermal and magnetic trips and contactor.

3.4 Control components

Various ranges of modular style control components are commercially available. The contactor is the basic unit which is designed to accept plug-in auxiliary contact modules, thermal overload relays and mechanical interlock kits, as illustrated in Fig. 3.5. Contactors can be snap-fitted on to a support rail. Control components are designed to IEC standards for heating, lighting and motor control applications.

A contactor is a switching device controlled by an electromagnet which can be used for power control applications. Figure 3.6 illustrates the relationship between the graphical symbol of a three-pole contactor and the connection terminals on the device. The graphical symbol consists of a contactor coil, mains contacts and a single auxiliary contact (13–14). The circuit between the mains supply and load is completed when the coil of the electromagnet is energized by the control circuit. The auxiliary contact(s) of a contactor are either integral to the device or a module which clips onto the top or side of the device. Auxiliary contacts are used in the design of a control circuit.

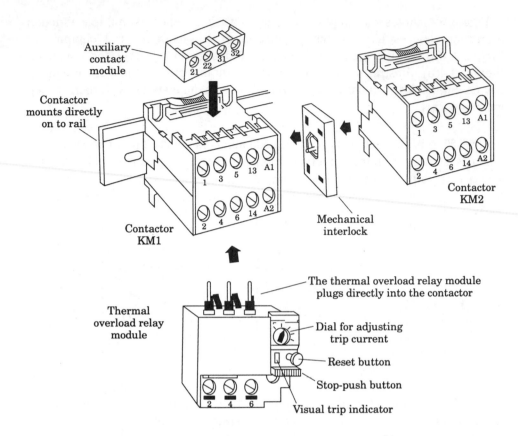

Figure 3.5   Control gear components.

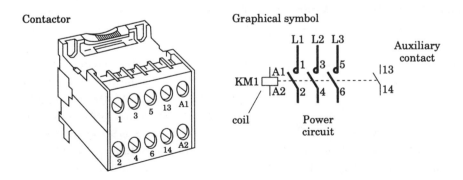

Figure 3.6   Contactor.

Commercially available contactors are of two types, namely a.c. operated contactors and d.c. operated contactors. In a.c. operated contactors the coil is energized using an a.c. supply. In d.c. operated contactors the coil of the contactor is operated with a d.c. supply. Low-voltage d.c. operated contactors are suited for use with PLC and other types of computer control systems. It is recommended that PLC control circuits are designed using a low-voltage supply such as 24 V d.c.

The convention for labelling the mains contacts of a contactor (and similar devices such as an isolator) is such that odd numbers are used to reference the top contacts, starting from the top left position. Numbering proceeds from left to right going from top to bottom. For example, the numbers 1,3,5 identify the top terminals and the numbers 2,4,6 identify the bottom terminals of the three-phase contactor shown in Figs. 3.5 and 3.6.

The convention for labelling auxiliary contacts is such that two digits are used. The first identifies the component position in the circuit diagram. The second digit indicates the function of the auxiliary contact as defined in Table 3.2. In Fig. 3.6, the reference notation 13 and 14 represents the normally open auxiliary contact associated with contactor KM1.

An example of a contactor fitted with an overload relay module is shown in Fig. 3.7. The overload relay module has a pre-wired built-in control circuit which is implemented by connecting the module blocks together. The power circuits illustrate a direct-on-line (dol) starter for both a three-phase and a single-phase motor. Isolation protection is not shown but, as explained above, is required.

Table 3.2 Auxiliary contact referencing

Auxiliary contact reference digit	Operation indicated
1,2	Normally closed contact
3,4	Normally open contact
5,6	Normally closed with special operation*
6,7	Normally open with special operation*

*An example of a special operation would be a time delay.

A thermal overload relay provides protection against small and prolonged current overloads. The principle of operation of the thermal overload relay is based on passing load current through a coil fitted to a bimetallic element. An increase in load current above the operating value heats the bimetallic element, causing it to bend and open the trip contact labelled 95–96 in the control circuit. Short circuit protection is provided by using fuses or a circuit breaker connected on the power supply side of the power circuit.

It is important to emphasize that the manual control circuit based on push-buttons 1 (start) and 0 (stop) is incorporated within the contactor and thermal overload relay blocks. The control circuit is drawn separately to indicate its operation.

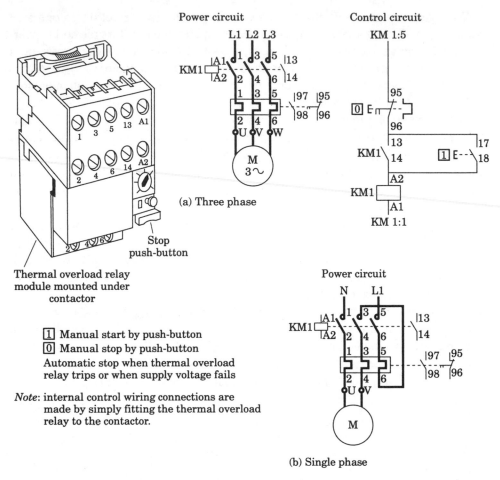

Figure 3.7 Thermal overload relay module for overcurrent protection in a motor circuit.

The supply is connected to the load when push-button 1 is pressed. The supply is disconnected by pressing push-button 0. The supply will also be disconnected when the thermal overload relay trips or when a supply voltage failure occurs.

The operation of the control circuit shown in Fig. 3.7 is as follows. When the start push-button 1 is pressed the contactor coil KM1 is energized and its associated auxiliary contact (13–14) closes. The auxiliary contact (13–14) holds the contactor coil on when the start push-button 1 is released. The auxiliary contact is referred to as a maintaining contact as it keeps the contactor coil KM1 electrically latched or held on. The latch can be broken by push-button 0 or by tripping the thermal overload relay contact (95–96) or by an interruption of the power supply. A supply voltage failure results in de-energizing the contactor coil and consequently the opening of the main and auxiliary contacts.

Two contactors can be electrically and mechanically interlocked to ensure that only one contactor can operate at any one time. Interlocked contactors are used for various applications such as reversing motor direction and mains transfer. A mechanical interlock is designed to fit between the two contactors. An electrical interlock is achieved using a circuit based on auxiliary contacts.[1] On a circuit diagram a dotted line and inverted triangle symbol indicate that the two contactors are interlocked.

3.5 Application example

The three-phase induction motor is widely used in automation applications such as conveyor belt operation. Injection brake modules[2] can be used for rapid braking of three-phase a.c. induction motors. The principle of operation is based on injecting a d.c. current into the stator windings producing a stationary magnetic field. This stationary field slows down the rotor and consequently produces a braking effect. Typical applications for injection braking include conveyor belts, mechanical handling equipment, rapid stopping of machinery to increase cycle or retooling time and braking of machinery which have a long run down time due to high inertia loads.

Fig. 3.8 shows a connection diagram for an injection brake module for use with direct-on-line (dol) starting of a three-phase induction motor. Braking commences when the 'brake-on' contact is closed. The brake-on contact can be a push-button or a PLC output relay contact.

3.6 Machine guard interlocking

A machine which poses a danger to the operator (e.g. in the form of moving parts) must be made safe by design processes such as providing a suitable guard or casing. Many machines have a protective casing and a guard in the form of a gate or door which can be opened for access. If a hazard is revealed by the opening of the guard, a safety interlock switch must be fitted which is actuated when the guard is moved. The function of the safety interlock switch is to isolate the power supply to the machine when the guard or access door is open. If there is an attempt to restart the machine whilst the guard is open the safety interlock switch should maintain the machine in a safe condition.

In practice, a safety gate guard will either activate an emergency stop or ensure that no hazardous machine movements occur within a zone. In the first case, a safety circuit can be configured with an emergency stop relay (see Section 3.7) by using positive drive safety switches in series with emergency stop buttons. The safety circuit must be hard wired being configured separately outside a PLC and have fail-safe redundancy to the level determined by a risk assessment for the machine.[6,11]

In the second case, a safety circuit can be configured using a unit which monitors safety gate limit switches. Such units are commercially available and are called safety gate monitors (see Fig 3.9).[4] These prevent machine movement while the gate is open by monitoring two separate inputs (i.e. the limit switches S1 and S2 in Fig. 3.9) to prove the position of the gate. To restart a machine requires that the inputs to the safety gate monitor unit be opened and closed in sequence to

Power circuit

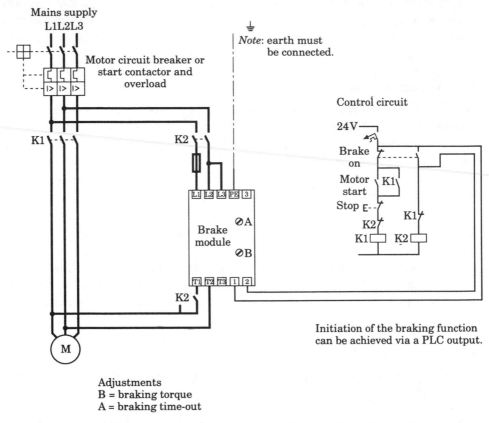

Figure 3.8   Application example: connection diagram for a three-phase motor dol starter with injection braking achieved via a PLC output.

allow the proper actuation of the relay. A safety gate monitor circuit must be hard wired and have a fail safe redundancy to the level determined by a risk assessment for the machine.[4,6,11]

A positive drive switch is one in which the actuator acts directly on the switching element so that the switching function does not rely on springs.[6] Positive driven switches are used in safety circuits because the switching function relies on forced disconnection. Likewise, when a guard door is opened its movement should physically pull apart (i.e. force disconnect) the safety contacts. The forced disconnection of the safety contacts by the movement of a guard is referred to as positive operation and overcomes sticking or welded contacts.[3] Safety interlock switches are commercially available for sliding, hinged or lift-off guards.

3.7   Emergency stop

A machine has to have a separate hard-wired emergency stop circuit and push-button to enable a user to stop its operation. An emergency stop must not depend on software or electronic logic and should not be connected through a PLC.

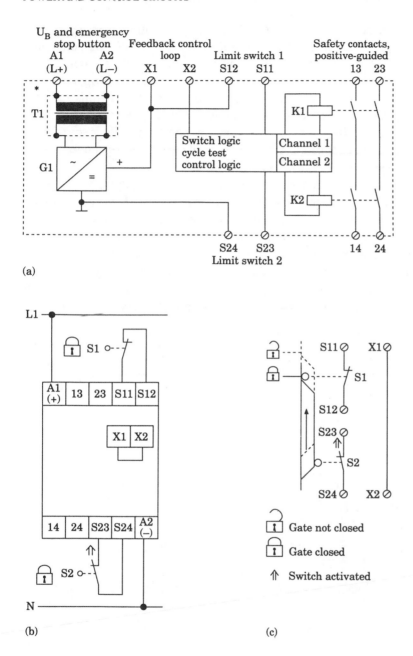

Figure 3.9 Safety gate monitor: (a) simplified internal circuit of safety gate monitoring relay, (b) connection diagram with safety gate shown in closed position and (c) safety gate with two forced contact limit switches. *Source*: RS Components Ltd data sheet B13113, March 1992.

Safety relays are commercially available for implementing emergency stop circuits. An example of an emergency stop safety relay (i.e. RS/Pliz emergency stop relay) which incorporates redundancy and cross-monitoring is shown in Fig. 3.10.[5] Redundancy is the duplication of control circuits which will operate if the primary circuit fails. Cross-monitoring ensures that if one circuit fails the fault will be detected and a reset cannot be implemented. This means that a machine is prevented from starting if a fault is detected on power-up. The provision of redundancy and cross-monitoring is required to fulfil current European legislation on machinery safety.[6]

The operation of the safety relay shown in Fig. 3.10 is as follows. When power is supplied to the system, K3 will energise and relays K1 and K2 are electrically latched on. The circuit based around relay K2 is the duplicate circuit providing redundancy. The normally closed contacts of K1 and K2 break the supply to K3 so that its contacts return to their initial states. At this stage, the positive guided safety contacts provide a route for connecting circuits. When the emergency stop is operated the safety contacts are switched off causing the instant removal of power to all connected devices. If one relay sticks then the safety contacts are still switched

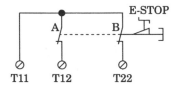

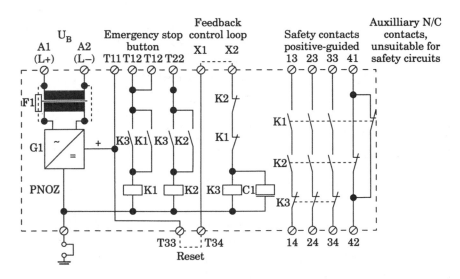

Figure 3.10   Emergency stop relay. *Sources*: RS Components Ltd data sheet D19240, 'Safety relays', March 1995 and RS Components Ltd data sheet B13450, 'Emergency stop unit Pilz UK Ltd', July 1992.

off. In this case, the system cannot be reset because the circuit to enable K3 is now broken. Positive guidance ensures that should a normally open contact on K1 (or K2) weld then the normally closed contact cannot be re-made on de-energization of the relay.

Note that the emergency stop relay, by incorporating an in-built step-down transformer and rectifier circuit, provides a low-voltage d.c. supply for connecting the emergency stop push-button. It is recommended that low voltage is used in control circuit implementation. The N/C auxiliary contact should not be used in the design of safety circuits but can be used by a PLC for monitoring the status of the emergency stop.[6]

3.8 Installation of automation equipment

The electrical installation of automation equipment should be undertaken by users who appreciate the safety aspects of the particular application. The onus is on the designer to create a machine which is both mechanically and electrically safe. Manufacturers and users of machinery are bound by The Supply of Machinery (Safety) Regulations 1992[7] which came into force on the 1 January 1993. This book does not cover safety-related system design but a practical overview is given in the RS Components data sheets.[3,6] European Standards relating to safety are referenced.[8–15]

References

1. *Practical Aspects of Electric Motor Controls*, Telemecanique Technical Manual, 1986.
2. 'Injection brake modules', RS Components data sheet B10120, March 1991.
3. 'Machine guard interlocking', RS Components data sheet B16718, November 1993.
4. 'Safety gate monitor', RS Components data sheet B13113, March 1992.
5. 'Emergency stop unit Pilz UK Ltd', RS Components data sheet B13450, July 1992.
6. 'Safety relays: a guide to the European machine safety directive', RS Components data sheet D19240, March 1995.
7. *The Supply of Machinery (Safety) Regulations 1992*, No. 3073, HMSO, London (Reprinted 1995).
8. *Safety of Machinery—Basic Concepts and General Principles of Design, Part 1: Basic Terminology, Methodology*, BS EN 292-1, 1991.
9. *Safety of Machinery—Basic Concepts and General Principles of Design, Part 2: Technical Principles and Specifications*, BS EN 292-2, 1991.
10. *Safety of Machinery—Rules for Drafting and Presentation of Safety Standards*, BS EN 414, 1992.
11. *Safety of Machinery—Risk Assessment*, PR EN 1050, 1993.
12. *Safety of Machinery—Electrical Equipment of Industrial Machines*, BS EN 60204, 1993.

13. *Safety of Machinery—Safety Related Parts of Control Systems*, PR EN 954-1,1992.
14. *Safety of Machinery—Emergency Stop Equipment, Functional Aspects and Principles for Design*, BS EN 418, 1992.
15. *Specifications for Low Voltage Switch Gear and Control Gear, Part 5: Control Circuit Devices and Switching Elements*, BS EN 60947-5-1, 1992.

4
Input devices

This chapter provides a brief overview of sensing devices used in manufacturing processes. Sensing devices with a digital output can be connected directly to the digital input port of a PLC. Sensors which produce an analogue voltage signal are connected using an analogue to digital converter.

4.1 Digital devices

The operation of a PLC may be based on signals received from digital switching devices which detect an event occurring. For example, the presence of an object on a conveyor belt can be detected using a proximity switch or a photoelectric detector. Some commonly used digital switching devices are described below.

4.1.1 PRESSURE AND TEMPERATURE SWITCHES

Pressure switches monitor pressure and have a contact which changes state when a pre-set pressure threshold value is reached. When the pressure level falls below the threshold value the contact resumes its initial position. Pressure switches are used in pneumatic and hydraulic applications.

Thermostats monitor temperature and have a contact which changes state when a pre-set temperature threshold value is reached. When the temperature value falls below the threshold value the contact resumes its initial position. Thermostats are used in a wide range of applications including the temperature monitoring and control of machines and heating installations. Many thermostats incorporate some hysteresis, which means that they switch on and off at different temperatures. This is referred to as differential regulation between two threshold values.

4.1.2 PROXIMITY SWITCHES

Proximity switches are used for non-contact object detection. They integrate a sensing element and a transistor switch output circuit. The output changes state when the object is in the vicinity of the sensor element. Proximity switches are classified as either inductive or capacitive, which relates to the type of sensing technology used. Inductive types are suitable for detecting ferrous and non-ferrous

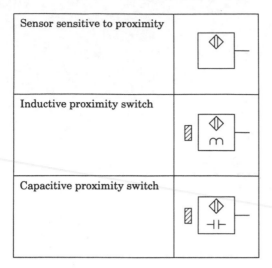

Sensor sensitive to proximity	
Inductive proximity switch	
Capacitive proximity switch	

Figure 4.1   The IEC graphical symbols for proximity switches.

metals. Capacitive types will sense the presence of almost any material. The IEC symbols used to represent proximity switches are shown in Fig. 4.1.

The principle of operation of an inductive proximity switch is based on the generation of an alternating magnetic field which is affected by a metal object passing within its range. A typical device would have a sensing range of 2 mm and a switching speed of 800 Hz. The principle of operation of a capacitive device is based on generating an electric field which is modified by any object passing within its range. A typical device would have a sensing range of 1–10 mm (adjustable) and a switching speed of 400 Hz.

The d.c. switching proximity devices are available with NPN open collector and PNP open collector transistor output stages. Figure 4.2 illustrates how to connect an NPN transistor output to a PLC sink input circuit and a PNP transistor output to a PLC source input circuit.

4.1.3   PHOTOELECTRIC SWITCHES

Photoelectric devices consist of a light source and a photo-receiver incorporating a transistor switch circuit. PNP and NPN transistor output devices are available. The ways in which the light source and photo-receiver can be set up to detect objects are illustrated in Fig. 4.3 on page 42. Photoelectric switches are categorized as being either through beam, mirror reflection, retro-reflective or diffuse reflective.

The through-beam type of photoelectric system has a separate transmitter and receiver. An object is detected when it breaks the light beam. The through-beam photoelectric detector can be used for long-range sensing. A typical device would have a sensing range of 8000 mm.

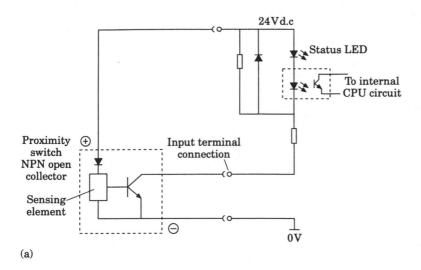

(a)

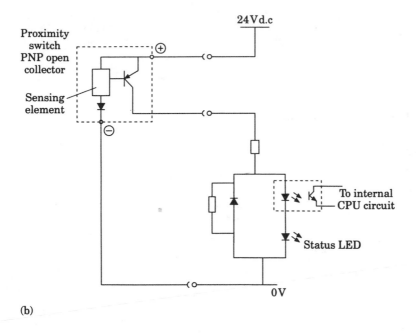

(b)

Figure 4.2 Connecting NPN and PNP transistor output proximity switches: (a) sink input circuit connection and (b) source input circuit connection.

A plane surface mirror can be used to reflect back the transmitted light beam to the receiver. In this case, the object is detected when the reflected beam is broken. With a plane mirror reflector, the transmitter and receiver must be mounted such that the angle of incidence equals the angle of reflection.

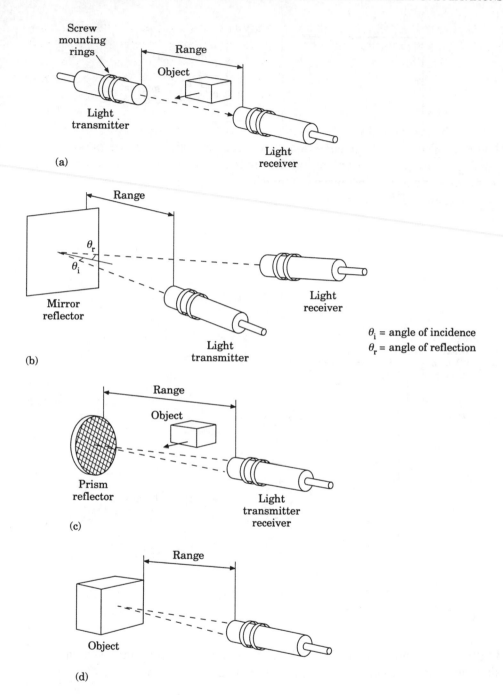

Figure 4.3 Photoelectric switch arrangements; (a) through beam, (b) mirror reflection, (c) retro-reflection and (d) diffuse reflection.

The retro-reflective type of photoelectric switch uses a special type of reflector which returns transmitted incident light back in the same direction from which it was sent. This allows the transmitter and receiver to be incorporated in the same housing. With proper alignment of the photoelectric device to the reflector a typical device would have a sensing range of 2000 mm.

If a parallel beam of light is incident on a sheet of paper the light is reflected in all directions because the surface is not perfectly smooth like that of a mirror. This is an example of diffuse reflection. A diffuse reflective type of photoelectric switch contains a transmitter and receiver in the same housing and switches when the diffuse reflection level exceeds a threshold value. A typical device would have a sensing range of 100 mm.

4.1.4   OPTICAL ENCODERS

Optical encoders convert either translation or rotary displacement into digital information. They operate by using a grating which moves between a light source and detector. There are two forms of optical encoder, namely incremental and absolute. Incremental encoders produce a pulse for each resolvable change in position. A relative measurement of position can be made by counting pulses from some reference point. Absolute encoders produce a unique coded number for each resolvable position.

An example of an incremental shaft encoder is shown in Fig. 4.4. The grating is a disc with a single concentric track of evenly spaced opaque and transparent regions. When light passes through a transparent region of the grating an output is obtained from the photo-detector. When the shaft is turned, the disc rotates and chops the light beam so that the photo-detector produces a series of electrical pulses. Position is measured by counting the number of pulses generated by the photo-detector as the disc rotates from a reference point. By incorporating two tracks shifted by a quarter-cycle relative to one another and two photo-detectors it is possible to produce two signals from which the direction of rotation can be determined by noting which signal rises first. If the shaft of an incremental encoder is driven by a motor the pulse rate is proportional to the motor's speed.

The disc of an absolute shaft encoder has several concentric tracks, with each track having an independent light source and photo-detector. With this arrangement a unique binary or Gray coded number can be produced for every shaft position. The Gray code changes by a single bit between successive positions (see Fig. 4.4). Table 4.1 on page 45 compares the Gray code with binary code. Resolution of an absolute encoder depends on the number of tracks (bits). For example, a 3-bit binary output produces 2^3 or 8 resolvable positions.

Linear encoders use a grating in the form of a slide to make a measurement of linear movement. They may be of incremental or absolute type. Rotary and linear absolute encoders read actual position and unlike incremental encoders they do not need a counting device.

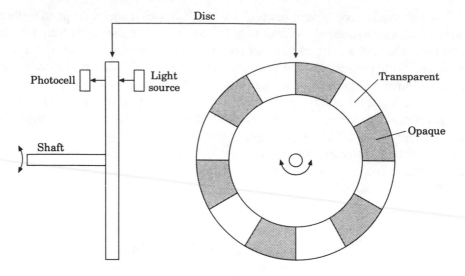

Incremental shaft encoder

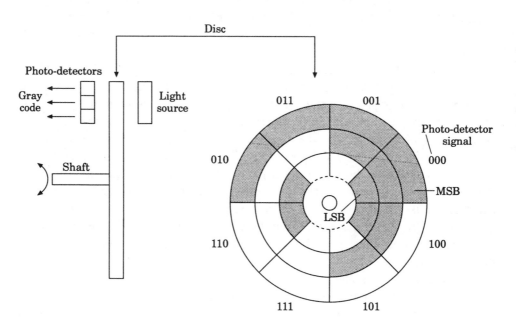

Absolute 3-bit Gray-coded shaft encoder

Figure 4.4   Incremental and absolute shaft encoders.

4.2   Analogue devices

Many applications involve monitoring analogue signals representing the variation of a physical quantity. For example, the variation of displacement can be measured using a linear variable differential transformer (LVDT) which converts a position displacement into a voltage signal.

Table 4.1 Binary and Gray codes

Binary	Gray
0000	0000
0001	0001
0010	0011
0011	0010
0100	0110
0101	0111
0110	0101
0111	0100

A typical PLC analogue input unit will accept either voltage or current signals (see Chapter 2). Current input ranges for data acquisition are commonly configured to accept 4–20 mA or 0–20 mA signals. Voltage input ranges can be unipolar (0–5 V for example) or bipolar (± 5 V for example). The input range is selected (usually via a jumper connection) to match the sensor signal variation. Some examples of analogue voltage sensors are described below.

4.2.1. LINEAR POTENTIOMETER

The linear potentiometer is used for measuring position and displacement. Modern devices consist of a printed circuit linear resistive track along which a slider makes contact (see Fig. 4.5). The slider is mechanically linked to the movement being measured. The track of resistive material is connected across a d.c. supply V. Assuming that the resistance is distributed linearly along the length L of the track the output voltage V_o from the slider can be written as

$$V_o = \frac{x_i}{L} V = K x_i$$

where

x_i = position of the slider
K = volts per unit length

The output voltage is directly proportional to the position of the slider along the linear track.

4.2.2 LINEAR VARIABLE DIFFERENTIAL TRANSFORMER (LVDT)

The linear variable differential transformer, or LVDT, is a displacement transducer. It consists of a nickel–iron rod which is free to move through primary and secondary coils. The basic arrangement is illustrated in Fig. 4.6.

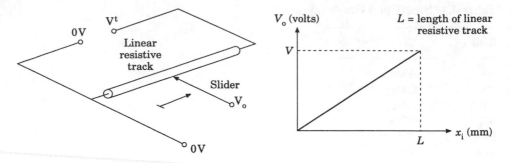

Figure 4.5   Linear potentiometer.

The primary coil is fed with alternating current so that voltages are induced in the two halves of the secondary coil. Moving the rod up and down changes the phase and voltage in the secondary windings. The output voltage versus core displacement characteristic in Fig. 4.6 shows that the phase of the output (secondary winding) relative to the input (primary winding) changes by 180° as the core is moved through the central position. Consequently, a phase detector is required to obtain an output for each core position.

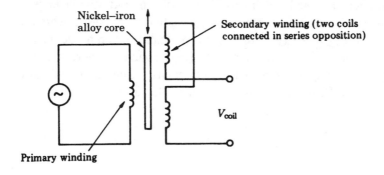

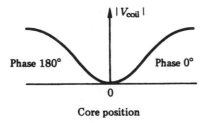

Figure 4.6   Linear variable differential transformer (LVDT).

4.2.3   TACHOGENERATOR

When the shaft of a permanent magnet d.c. motor is driven mechanically an output voltage is produced whose magnitude is proportional to the speed of rotation and polarity depends on the direction of rotation. A permanent magnet d.c. motor used for speed sensing rather than a machine for producing power is called a tacho-generator. Low-pass filtering is required to reduce the ripple voltage super-imposed on the output.

4.2.4   TEMPERATURE SENSORS

A thermocouple is a device that converts temperature into a voltage. It consists of two dissimilar wires which are arranged as shown in Fig. 4.7. Voltage is produced by thermoelectric effects as the hot junction is heated. Thermocouple types are designated a letter which indicates the types of metals used in the thermocouple junction (see Fig. 4.7).

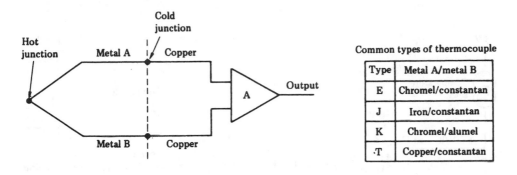

Common types of thermocouple

Type	Metal A/metal B
E	Chromel/constantan
J	Iron/constantan
K	Chromel/alumel
·T	Copper/constantan

Figure 4.7   Thermocouple.

Thermocouples are non-linear devices which means that their output voltage is not proportional to temperature. A thermocouple is supplied with a calibration table of output voltages versus temperature. The output voltage produced by a thermocouple is at the millivolt level and needs to be amplified before it can be fed into an ADC unit. A simple thermocouple amplifier circuit based on the 741 operational amplifier is shown in Fig. 4.8.

Semiconductor temperature sensors (e.g. RS590) are commercially available which can generate current as temperature is increased. A typical device would have a temperature coefficient of 1 μA/K and operate in the temperature range of 218–403 K.

Figure 4.9 shows a temperature switch circuit based on a semiconductor temperature sensor. The circuit converts sensor current into voltage using the resistor R1. This voltage is amplified by the operational amplifier circuit and fed to a comparator. The comparator produces an output signal when the temperature voltage input is equal to or greater than the threshold level set by V_{set}. In this case the comparator output level turns on the transistor switch based on T1. The

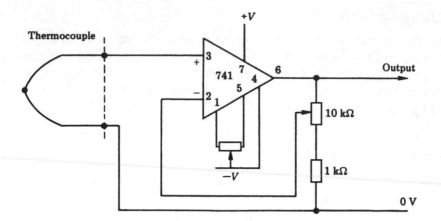

Figure 4.8   Basic thermocouple amplifier circuit.

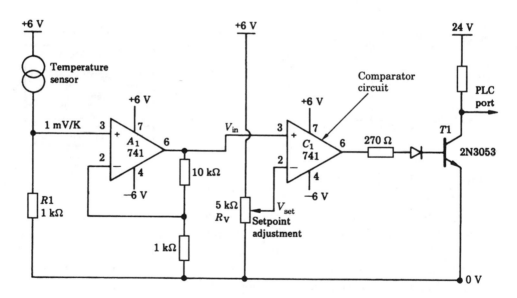

Figure 4.9   Temperature switch circuit.

transistor switch T1 allows 24 V to be applied to the PLC digital input point. Note that the output to the PLC port is held on until a base current flows in T1.

4.2.5   STRAIN GAUGE

A strain gauge is a transducer whose principle of operation is based on the variation of resistance with dimensional displacement. A strain gauge is bonded to a surface of a mechanical element (e.g. a bar) in which strain is to be measured.

Provided that the variation in length under loaded conditions is along the gauge-sensitive axis an increase in load causes an increase in gauge resistance. The relationship between the change in resistance ($\Delta R/R$) and the corresponding change in strain (i.e. the length change ($\Delta L/L$) is

$$G = \frac{\Delta R/R}{\Delta L/L}$$

where G is called the gauge factor. The gauge factor is about 2 for wire element metal alloy strain gauges and about 100 for semiconductor strain gauges.

A strain gauge is normally connected in a Wheatstone bridge arrangement as shown in Fig. 4.10. The bridge is balanced under no load conditions so that any change in resistance due to loading unbalances the bridge and a signal is detected. A dummy gauge can be connected in the bridge to compensate for the change in resistance of the gauge due to temperature variations. An instrumentation amplifier is used to feed the balance signal to an ADC input point.

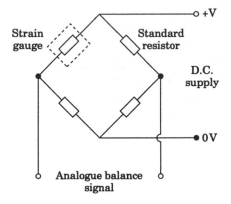

Figure 4.10   Strain gauge bridge.

4.3   Basic interfacing techniques

The voltage signal from a sensor will need to be matched to the specified voltage range and internal resistance of an ADC input port. A typical input port specification is an internal resistance of 200 kΩ for an input range of 0–5 V.

Matching the signal source voltage level to that of the input specification may require reducing or amplifying the voltage of the signal source. If the signal source voltage ranges between −2.5 and +2.5 V (i.e. a bipolar signal) it will need to be converted into the range 0–5 V (i.e. a unipolar signal) before connecting to an input specified at 0–5V. It is also necessary to ensure that the resistance (i.e. impedance) of the signal source is less than or equal to that of the input specification. An impedance changing circuit (e.g. emitter follower) may be required.

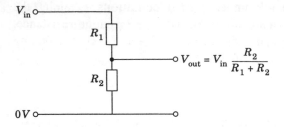

Figure 4.11   Voltage divider.

A voltage (potential) divider, as shown in Fig. 4.11, can be used to reduce the voltage level of a transducer signal. The output voltage V_{out} is always less than the input voltage V_{in} and is given by

$$V_{out} = V_{in} \frac{R_2}{R_1 + R_2}$$

Small signal amplifiers are used to amplify voltages at the microvolt to millivolt level. Figure 4.12 shows circuits for inverting and non-inverting amplifiers based around a 741 operational amplifier. The ratio of the output voltage V_{out} to the input voltage V_{in} is called the gain of the amplifier. The inverting amplifier has a gain of $-R_2/R_1$. The non-inverting amplifier has a gain of $(R_2+R_1)/R_1$. The resistance R_i in these circuits ensures that both inputs to the operational amplifier see the same resistance to 0 V and is calculated using

$$R_i = \frac{R_1 R_2}{R_1 + R_2}$$

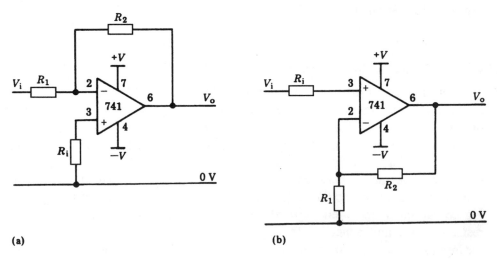

(a)   (b)

Figure 4.12   (a) Inverting and (b) non-inverting amplifiers.

A summing amplifier can be used to convert a bipolar signal voltage into a unipolar voltage. Figure 4.13 shows a circuit for a summing amplifier which makes use of a 741 operational amplifier. The output voltage V_{out} is given by

$$V_{out} = -(V_1 + V_2)$$

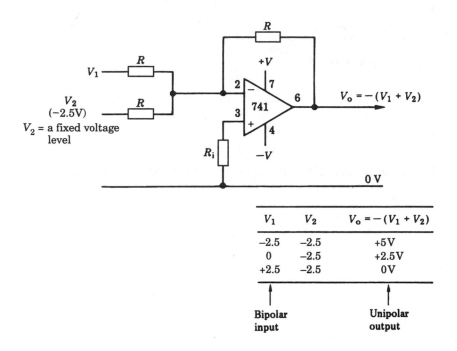

V_1	V_2	$V_o = -(V_1 + V_2)$
−2.5	−2.5	+5V
0	−2.5	+2.5V
+2.5	−2.5	0V

Bipolar input → Unipolar output

Figure 4.13   Summing amplifier.

If the summing amplifier is to convert a bipolar signal voltage into a unipolar voltage, one of the inputs must be held at an appropriate negative voltage. For example, if V_2 is held at −2.5 V, then summing action ensures that the bipolar voltage range −2.5 to +2.5 V is converted into the unipolar voltage range 5–0 V.

The emitter follower and voltage follower (a unity gain non-inverting amplifier) are used as impedance matching circuits. Figure 4.14 shows circuits for the emitter follower and voltage follower. In these circuits the output voltage follows the input voltage (e.g. from the signal source). The main characteristic of both circuits is that the input impedance is high and the output impedance is low. Consequently, they can be used as a buffer between a sensor and ADC port to ensure that the impedance matching criteria is met. Buffer amplifiers are also used for interfacing low current sensors.

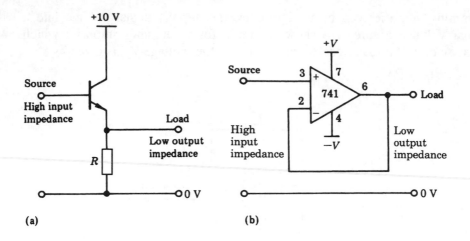

Figure 4.14 Impedance changing circuits: (a) emitter follower and (b) voltage follower.

5

Programming methods

5.1 The IEC 1131-3 programming languages standard

This chapter provides an overview of the recently developed programming language standard for PLCs, namely the IEC 1131-3.[1] The standard published by the International Electrotechnical Commission (IEC) in 1993 builds upon programming methods in use today and developed by different PLC manufacturers. It is a major step forward in establishing standardized forms for the following five PLC programming languages:

- *Structured text (ST)* A high-level text-based language similar to Pascal for developing well structured control software
- *Instruction list (IL)* A low-level instruction list language based on mnemonics common to current mainstream PLCs
- *Ladder diagram (LD)* A graphical programming language evolved from electrical relay logic circuit methods and used by all mainstream PLC manufacturers
- *Function block diagram (FBD)* A graphical programming language based on function blocks which can be re-used in different parts of an application
- *Sequential function chart (SFC)* A graphical language for developing control sequence programs that are time and event driven and which is based on Grafcet[2]

The standard promotes an open systems environment to enhance integration between different PLC systems such as the possibility of being able to port software between two IEC 1131-3 compliant systems. At present, any PLC system can claim to be IEC 1131-3 compliant by implementing one or two languages. PLC manufacturers are offering the designer a choice of programming languages based on the IEC standard so that a language suited to the application can be used. It is not always possible to directly translate from one language to another. For example, the structured text constructs IF … THEN, CASE, WHILE, REPEAT are not readily represented graphically within the ladder and function block languages.

The standard uses the terms:

1. *Configuration* to mean a program that specifies the particular arrangement of PLC hardware on to which application software is mapped.
2. *Resource* to mean a processing facility able to execute an IEC program, e.g. a processor board within a multiple processor PLC.
3. *Program organization unit (POU)* to mean software elements, i.e. programs, function blocks and functions. In general, a program can be thought of as a number of interconnected function blocks which provide the required control operation.
4. *Task* to mean the scheduling of software elements, e.g. POUs. A program (or function block) can be associated with a task such that two or more tasks can be run by sharing processor time among them (e.g. multitasking). The IEC standard allows for tasks that can be executed at different rates according to the needs of the application.

5.2   Common elements

The IEC standard (see Tables 5.1 and 5.2) defines a number of common programming features shared by all five languages. A brief overview of some essential common programming features is given below.

5.2.1.   IDENTIFIERS

The names that are used to reference variables, labels, function blocks, programs, etc., are called 'identifers'. The standard specifies that at least six characters of uniqueness shall be supported in IEC systems. The first character must not be a digit but otherwise the name can be made up of letters, digits and a single underscore. An example of an identifier is the variable name 'input1'.

5.2.2.   DATA TYPES

The IEC standard allows the programmer to use and define a wide variety of data types including different sizes of integers and floating point (e.g. real) values. String, date and time data types are also defined. Table 5.1 lists the main IEC data types. High-level language structures such as arrays and records are defined for use in structured text.

The time and date data types are useful as many control applications require the date and time of events. Examples of time literals (i.e. numerical constants) are:

T#5m5s5ms (*Short form which defines 5 minutes, 5 seconds and 5 milliseconds*)

TIME#5m_5s_5ms   (*Long form but otherwise as above*)

The letters d for day, h for hour, m for minutes, s for seconds and ms for milliseconds are used.

5.2.3   VARIABLES

Variables provide a means of identifying data objects whose contents may change.

Variables can be of any data type such as real, integer or Boolean. A variable can be assigned a value in the following way:

pressure:=10;   (*Assigns the value '10' to the variable pressure*)

Table 5.1 IEC standard data types

IEC standard data type	Description
INT	Integer or whole number (−32768 to 32767)
SINT	Short integer (−128 to +127)
DINT	Double integer (-2^{31} to $+2^{31}-1$)
LINT	Long integer (-2^{63} to $+2^{63}-1$)
USINT	Unsigned short integer (0 to 255)
UINT	Unsigned integer (0 to $2^{16}-1$)
UDINT	Unsigned double integer (0 to $2^{32}-1$)
ULINT	Unsigned long integer (0 to $2^{64}-1$)
REAL	Real or floating point number (-10^{-38} to $+10^{38}$)
LREAL	Long real number (-10^{-308} to $+10^{308}$)
TIME	Time duration in days(d), hours(h), minutes(m), seconds(s) and milliseconds(ms)
DATE	Calendar date
TIME_OF_DAY (or TOD)	Time of day as obtained by a system real-time clock
DATE_AND_TIME (or DT)	Date and time of day
STRING	Character string to store textual information
BOOL	Bit string of one bit; e.g. can have one of two states: TRUE (1) or FALSE(0)
BYTE	8-bit binary string
WORD	16-bit binary string
DWORD	32-bit binary string
LWORD	64-bit binary string
F_EDGE	Falling edge
R_EDGE	Rising edge
ANY	Overloaded variables in functions or function blocks

The IEC standard allows PLC memory locations to be referenced directly and the % character is used for this purpose. As discussed in Chapter 2, PLC memory is organized into three main regions, namely input image memory (I), output image memory (Q) and internal memory (M). A memory location is identified by selecting options from the following list:

% (I or Q or M) (X or B or W or D or L) (one or more numeric fields)

The second letter codes are X for bit, B for byte, W for word, D for double word (32 bits) and L for long word (64 bits). If none of these are used, memory bit organization is assumed. An example is:

%IX22   (*Input memory bit 22*)

Global variables (i.e. those declared for use throughout a program) and local variables (e.g. those declared for use within a function block) are supported by the standard.

Table 5.2 IEC standard keywords

IEC standard keyword	Description
ACTION/END_ACTION	Used to declare the start and end of an action block. SFC action blocks can have the following qualifiers: N=none, S=set, R=reset, L=limited time action, D=delayed action, P=pulse action, SD=stored and delayed, DS=delayed and stored and SL=stored time limited
ARRAY OF	Declares an array data type
AT	Used to equate a particular variable with a memory location. For example, CR1 AT %QX400
CASE OF/ELSE/END_CASE	Case construct in structured text
CONFIGURATION/ END_CONFIGURATION	Used to declare the start and end of a configuration file

CONSTANT	Used to declare values that will remain constant. For example,
	    VAR CONSTANT
	    PI : REAL := 3.14
	    END_VAR

EN/ENO	Enable execution control used with functions in the ladder diagram language
EXIT	Used to exit a loop in structured text
FOR/TO/BY/DO/END_FOR	For loop construct in structured text
FUNCTION/END_FUNCTION	Used to declare the start and end of a function. A function is different to a function block as it produces a single-value result with no internal storage. A function would be used to produce, for example, Y:= SIN(X)*SIN(X)
FUNCTION_BLOCK/ END_FUNCTION_BLOCK	Used to declare the start and end of a function block

Table 5.2 IEC standard keywords *(continued)*

IEC standard keyword	Description
IF/THEN/ELSIF/ELSE/END_IF	Used to define the IF/THEN/ELSE construct in structured text
INITIAL_STEP/END_STEP	Used to declare an initial step
PROGRAM/ END_PROGRAM	Used to declare the start and end of a program
READ_ONLY/READ_WRITE	Variable attributes
REPEAT/UNTIL_END_REPEAT	Used to define the repeat construct in structured text
RESOURCE/ON/ END_RESOURCE	Used to define the start and end of a resource configuration

RETAIN	Attribute which specifies that the state of a variable is to be retained during PLC power interruption. For example,
	    VAR_OUT RETAIN
	    Status : INT
	    END_VAR

RETURN	Used to prematurely exit a function or function block
STEP/END_STEP	Used to define the start and end of a step (see SFC language)
STRUCT/END_STRUCT	Used to declare a record type structure
TASK	Used to declare a task
TRANSITION/FROM/TO END_TRANSITION	Used to define the start and end of a transition (see SFC)
TYPE/END_TYPE	Used to define the start and end of a derived data type
VAR/END_VAR	Used to declare a set of internal variables
VAR_INPUT/END_VAR	Used to declare a set of input variables
VAR_IN_OUT/END_VAR	Used to declare a set of input and output variables
VAR_OUTPUT/END_VAR	Used to declare a set of output variables
VAR_EXTERNAL/END_VAR	Used to declare a set of external variables
VAR_ACCESS/END_VAR	Used to declare a set of variables which may be accessed by other extrinsic configurations
VAR_GLOBAL/END_GLOBAL	Used to declare a set of global variables which can be accessed by all elements within a program
WHILE/DO/END_WHILE	Used to define the WHILE construct in structured text
WITH	Used to associate a program or function block with a task as part of a configuration

Other reserved names include BOOL, INT, REAL, TRUE, FALSE, NOT, OR, AND, XOR, RS, TON, LD, ST, ADD, MUL, DIV, SIN, COS, etc.

5.2.4.   LANGUAGE KEYWORDS

Language keywords define different constructs or the beginning and end of a particular software element. For example, the keywords PROGRAM, END_PROGRAM define the start and end of a program unit. The IEC language keywords are outlined in Table 5.2. For a full coverage of common elements refer to the standard.[1]

5.3   Textual languages

The textual languages defined in the IEC standard are structured text (ST) and instruction list (IL). These languages can be used in conjunction with sequential function chart elements.

5.3.1   STRUCTURED TEXT (ST)

Structured text (ST) is a high-level language for control which uses constructs such as IF ... THEN, CASE, FOR, WHILE, REPEAT and the object-orientated feature of encapsulation (e.g. the restriction of external access or hiding of software internals within program functions). Comparisons can be drawn with Pascal, which structured text resembles.

In ST, program structure takes has the general form:

```
PROGRAM ProgName
  VAR_INPUT
    (*list each input variable and its data type*)
  END_VAR
  VAR_OUTPUT
    (*list each output variable and its data type*)
  END_VAR
  VAR
    (*list each internal variable and function block
    used within the program*)
  END_VAR
    (*main program body*)
END_PROGRAM
```

Statements are separated using a semicolon. Comments to aid readability of the program are enclosed between (*and*).

Reusable function blocks having internal storage can be defined using the keywords FUNCTION_BLOCK and END_FUNCTION_BLOCK. A function block written with the ST language can be used by other IEC 1131-3 languages. Function block structure is similar to program structure, namely:

```
FUNCTION_BLOCK FuncName
  VAR_INPUT
    (*list  function  input  variables  and  their  data
    type*)
  END_VAR
  VAR_OUTPUT
    (*list  function  output  variables  and  their  data
    type*)
  END_VAR
  VAR
    (*internal variable declarations*)
  END_VAR
    (*main function block body e.g. algorithm*)
END_FUNCTION_BLOCK
```

Operators common to modern high level languages are defined within the IEC standard. These include:

- Arithmetic operators for addition (ADD or +), multiplication (MUL or *), subtraction (SUB or −) and division (DIV or /)
- Equality operators such as greater than (GT or >), greater than or equal (GE or >=), equal (EQ or =), less than or equal (LE or <=), less than (LT or <) and NOT equal (NE or <>)
- Bitwise operators (AND, OR, XOR, NOT)

A set of numerical functions including SIN, COS, TAN, SQRT and LOG are also defined within the standard. Data type overloading can be applied to these functions which means that they can be used with different data types such as real and integer numbers.

The IF/THEN/ELSE construct has the form:

```
IF <Boolean expression is true> THEN
<statement>
ELSE
<statement>
END_IF
```

In structured text, the ELSE clause is optional and the <statement> can be replaced with a block of statements.

The CASE construct allows different statements to be selected and has the form:

```
CASE <integer expression> OF
<list> : <statements>
<list> : <statements>
...
ELSE
<statements>
END_CASE
```

In ST, <list> is a list of integer values to select the various case options.

Structured text has three kinds of looping statements: the FOR ... DO loop, the WHILE ... DO loop and the REPEAT ... UNTIL loop. The FOR ... DO construct has the form:

```
FOR <index> := <start> TO <finish> BY <increment> DO
    <statements>
END_FOR
```

The WHILE ... DO construct has the form:

```
WHILE <Boolean expression is true> DO
<statements>
END_WHILE
```

The REPEAT ... UNTIL construct has the form:

```
REPEAT
<statements>
UNTIL <Boolean expression is false>
END_REPEAT
```

5.3.2 INSTRUCTION LIST (IL)

The standard describes the use of a low-level instruction list (IL) language for programming a PLC. An instruction is made up of an operator followed by one or more operands. An operand is the quantity upon which an instruction operation is performed. Assembly language type operators are defined by the standard and these are listed in Table 5.3. Jump to label and call function operators are included. Labels can be assigned to addresses in the program. An example of an instruction list code fragment is given below:

Label	Operator	Operand	Comment
	LD	Count	(* Load the current count value into accumulator *)
Loop	SUB	1	(* Decrement by one *)
	NE	0	(* Test if not equal to zero *)
	JMP	Loop	(* Jump back to the label 'Loop' while not equal to 0 *)

Single letter modifiers can be used with some IL operators to change the meaning of the instruction. For example, the AND instruction operator can be modified to ANDN (i.e. AND NOT operator). Parenthesis or bracket modifiers allow instructions to be deferred in the same way as brackets are used in normal arithmetic expressions. For example, the following code fragment defers the 'greater than' comparison test until (A+B) has been calculated:

```
LD    X    (* Load X into the accumulator *)
GT(   A    (* Defer GT operation until A+B is calculated *)
ADD   B
)          (* Test if X GT (A+B) *)
```

Table 5.3 Instruction list (IL) operators

Instruction Operator	Modifiers	Operand type	Description
LD	N	Any	Load operand into accumulator (e.g. LD %IX1)
ST	N	Any	Store accumulator into operand (e.g. ST %QX2)
S		BOOL	Set operand to 1 (TRUE)
R		BOOL	Set operand to 0 (FALSE)
AND	N and ( )	BOOL	Logical AND
OR	N and ( )	BOOL	Logical OR
XOR	N and ( )	BOOL	Exclusive OR
ADD	( )	Any	Addition
SUB	( )	Any	Subtraction
MUL	( )	Any	Multiplication
DIV	( )	Any	Division
GT	( )	Any	Greater than comparison operator
GE	( )	Any	Greater than or equal to comparison operator
EQ	( )	Any	Equal to comparison operator
NE	( )	Any	Not equal to comparison operator
LE	( )	Any	Less than or equal to comparison operator
LT	( )	Any	Less than comparison operator
JMP	C,N	Label	Jump to label
CAL	C,N	Name	Call function block by name
RET	C,N		Return from function or function block

Modifiers are: N = NOT, ( ) = deferred execution and C = conditional execution. Instructions that can be used with any data type are said to be overloaded. The accumulator is a CPU data register.

Many PLC systems use a proprietary instruction set whereby each ladder rung is translated into mnemonic instructions and data. Examples of device-specific instruction sets are given in Appendices 3, 4 and 5.

5.4   Graphical languages

Graphical languages are widely used for programming PLCs as they are easy yet powerful tools for developing control applications. Computer-based graphical programming packages exist for designing ladder diagrams, function block diagrams and sequential function charts. These literally allow the user to draw a control network and provide a visual image of the solution to a particular control problem. The graphical languages defined in the IEC standard are ladder diagram

(LD) and function block diagram (FBD). Sequential function chart (SFC) elements are also defined and can be used in conjunction with either of these languages.

A circuit network can be considered as a set of interconnected graphical elements representing a control plan. Graphical languages are used to represent the flow of a conceptual quantity through a network. With a ladder diagram network the conceptual quantity is power flow and the direction of power flow is defined to be from left to right. With a function block diagram network the concept of signal flow is applied and the direction of signal flow is defined to be from the output side to the input side of connected function blocks or functions. With a sequential function chart the concept of activity flow is applied. Activity flow between SFC elements is from the bottom of a step through the appropriate transition to the top of the following step.

5.4.1   LADDER DIAGRAM (LD)

The ladder diagram method has been used throughout this book as it is the predominant programming method for developing small- and medium-scale PLC applications. The IEC standard is based on terminology and symbols as used by the majority of current PLC systems. The IEC standard ladder diagram graphical symbols are shown in Table 5.4. The standard includes symbols for SET and RESET coils which allow a variable to be latched on and then cleared at a later stage. Retentive coils (e.g. coils whose Boolean states need to be held during PLC power interruption) are also defined.

Function blocks having Boolean inputs and outputs can be connected within a ladder diagram. Functions used in a ladder diagram may have an execution enable (EN) input and an execution enable output (ENO). When EN is true the function is evaluated. When EN is false it remains inactive. A ENO is taken high on the successful completion of a function. It is possible to connect an ENO output to an EN of another function to provide execution control.

To transfer control from one part of a ladder to another requires using a jump construct and label identifier. However, the IEC does not recommend the use of a jump construct although many systems support its use. Ladder diagram examples are covered extensively in Chapters 6, 7 and 8.

5.4.2   FUNCTION BLOCK DIAGRAM (FBD)

The function block diagram (FBD) is a graphical language whereby programs are expressed as a set of interconnected function blocks. An analogy can be drawn with a system circuit diagram where connections represent signal flow paths. Each function block is represented as a rectangular block with inputs drawn so that they enter on the left side and outputs drawn so that they exit on the right side. Signal flow is from an output of one function block to the input of another function block. An example of a function block diagram is shown in Fig. 5.1.

Table 5.4 IEC standard ladder diagram symbols

IEC standard ladder graphical symbols	Description	IEC standard ladder graphical symbols	Description
—||—	Normally open contact	—(R)—	RESET coil: The coil variable is reset OFF (i.e. cleared) when the left-hand link is ON. The coil remains OFF until it is set using the SET coil
—|/|—	Normally closed contact	—(M)—	Retentive memory coil: Acts in the same way as a normal coil but its state is retained during a PLC power interruption
—|P|—	Positive transition-sensing contact: The condition of the right link is ON for one ladder rung evaluation when a change from 0 to 1 (i.e. positive rising edge) is sensed	—(SM)—	SET retentive memory coil: Acts in the same way as the SET coil but its state is retained during a PLC power interruption
—|N|—	Negative transition-sensing contact: The condition of the right link is ON for one ladder rung evaluation when a change from 1 to 0 (i.e. negative falling edge) is sensed	—(RM)—	RESET retentive memory coil: Acts in the same way as the RESET coil but its state is retained during a PLC power interruption
—( )—	Coil: The coil is set from the left-hand link	—(P)—	Positive transition sensing coil: The coil variable is set ON for one ladder rung evaluation when a change from 0 to 1 (i.e. positive rising edge) is sensed on the left-hand link
—(/)—	Negated coil: The negated coil is set to the opposite state to that of the left-hand link	—(N)—	Negative transition sensing coil: The coil variable is set ON for one ladder rung evaluation when a change from 1 to 0 (e.g. negative falling edge) is sensed on the left-hand link
—(S)—	SET coil: The coil variable is set to the ON state when the left-hand link is ON. The coil remains ON (i.e. latched) until it is reset using the RESET coil		

63

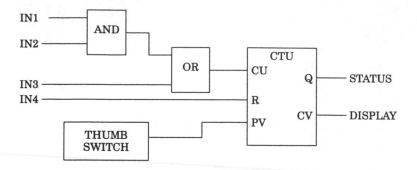

Figure 5.1   Example of a function block diagram (FBD).

The IEC standard defines a small number of standard function blocks which are the SR bistable, RS bistable, rising edge detector, falling edge detector, edge detecting inputs, up-counter, down-counter, up-down counter, pulse timer, on-delay timer, off-delay timer and real-time clock. These are shown in Table 5.5 on page 66–67. Other special purpose blocks such as PID and ramp blocks can be designed using the ST language.

The FBD language is suited to Boolean logic and continuous (e.g. closed-loop) control applications. Although it is possible to construct a 'jump' within an FBD using a label identifier, the IEC does not recommend its use. Constructs such as IF/THEN/ELSE used with the structured text language are difficult to represent graphically within an FBD. Consequently, direct translation between ST and FBD is not always possible.

5.5   Sequential function chart (SFC)

The IEC 1131-3 standard describes the use of a graphical sequencing language referred to as the sequential function chart. This is based on Grafcet, a graphical language for developing sequential control programs, and defined as a French national standard. Telemecanique PLCs use the Grafcet language but most major PLC systems provide an option for sequential programming (e.g. the Mitsubishi Stepladder discussed in Appendix 3). In fact, the IEC 1131-3 standard uses an existing IEC standard to describe a graphical language for control sequences.[2]

A sequence can be thought of as a series of steps that occur in a defined order. The sequential function chart language represents each step as a rectangular box which is associated with a control action. An action can be drawn as a rectangular box that is attached to a step. Every sequence starts with an initial step which is concerned with holding the system ready for operation. Connecting lines between steps have a horizontal bar representing a transition condition. When a transition condition becomes true, the step before the transition is deactivated and the step after the transition is activated. The main features of an SFC are illustrated in Fig. 5.2.

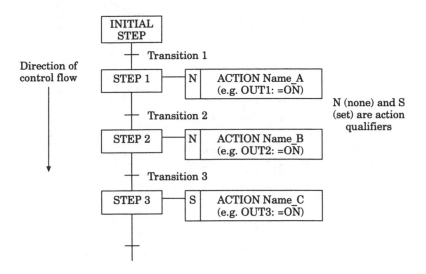

Figure 5.2   Sequential function chart.

Steps are always separated by a transition and every transition must have a Boolean condition which can be true. Every action and transition should be given a unique name within a program.

The software to implement transition conditions and actions can be written using any of the IEC languages. An example of a transition condition (Trans1) written in ST is

```
TRANSITION Trans1:
  := Startbutton AND Ready;
END_TRANSITION
```

The same transition is shown as a Ladder Diagram in Fig. 5.3 on page 68. A transition can involve a timer (e.g. wait for an elapsed time to occur), a counter or any other type of function block.

An example of a simple action written in ST is

```
ACTION Name_A
  OUT1:=ON;
END_ACTION
```

The action called 'Name_A' sets OUT1 to an ON state. The action will occur as a result of an associated step (e.g. STEP 1 in Fig. 5.2) being activated.

An action can have a qualifier which determines the way in which the action will be executed. An action qualifier is drawn as a rectangular box attached to the left hand side of an action as shown in Figure 5.2. Commonly used qualifiers are N (none), S (set) and R (reset). An N qualifier specifies that the action is executed only while the step is active. An S (set) qualifier specifies that the action is to be latched on. The step acts as a trigger for the latch and the action is said to be

Table 5.5 IEC standard function blocks

IEC standard function blocks	Label description
SR bistable	The SR bistable is a latch where the set (S1) input takes priority. S1 = set input of type BOOL R = reset input of type BOOL Q1 = latched output of type BOOL
RS bistable	The RS bistable is a latch where the reset (R1) input takes priority. S = set input of type BOOL R1 = reset input of type BOOL Q1 = latched output of type BOOL
Rising edge detector	The rising edge detector sets the output (Q) true on the rising edge of the input (CLK) for a duration of one function block execution. CLK = clock input of type BOOL Q = output of type BOOL
Falling edge detector	The falling edge detector sets the output (Q) true on the falling edge of the input (CLK) for a duration of one function block execution. CLK = clock input of type BOOL Q = output of type BOOL
Edge detecting inputs	The edge detecting inputs detector sets the output (Q) true when either edge inputs are true. A right-to-left arrow is used to indicate a falling edge (see CLK1). A left-to-right arrow is used to indicate a rising edge. In this example, Q is true when either CK1 detects a falling edge or when CLK2 detects a rising edge. CK1 = input (falling edge) of type BOOL CK2 = input (rising edge) of type BOOL Q = output of type BOOL
Up-counter	The up-counter counts the number of rising edges on the input CU until a pre-set value (PV) is reached. At this point, the output (Q) is set true and counting stops. The reset input (R) can be used to restart the counter action. CU = count-up input (BOOL) R = reset input (BOOL) PV = pre-set value (INT) Q = output (BOOL) CV = current count value (INT)
Down-counter	The down-counter counts down the number of rising edges on the input (CD) from a pre-set value until zero is reached. The output (CV) is initially loaded with the pre-set value and is decremented by one each time a rising edge is detected on the input (CD). CD = count down on rising edge input (BOOL) LD = load CV with PV and clear Q (BOOL), i.e. reset PV = pre-set value (INT) Q = output (BOOL) CV = current count value (INT)

Table 5.5 IEC standard function blocks (continued)

IEC standard function blocks	Label description
On-delay timer	The on-delay timer is used to delay the setting of the output Q high after the input IN is taken high. When IN is set true, the elapsed time ET increases and when it is equal to the pre-set time PT, the output Q goes high and ET is held. The output Q is reset false by taking IN low. If IN is taken low before the pre-set time PT occurs then Q remains low. IN = timer start input (BOOL) PT = pre-set 'wait' time (TIME) Q = delayed output (BOOL) ET = elapsed time (TIME)
Off-delay timer	The off-delay timer is used to hold the output for a given time duration PT after the input IN is taken low. IN = timer start input (BOOL) PT = pre-set 'hold period' (TIME) Q = held output (BOOL) ET = elapsed time (TIME)
Real-time clock	The real-time clock function allows the current date and time to be read once it has been pre-set to an initial date and time (e.g. when used for the first time). EN = enable input (BOOL) PDT = pre-set date and time (DT) Q = output (BOOL) CDT = current date and time (DATE_AND_TIME)

IEC standard function blocks	Label description
Up-down counter	The up-down counter counts the number of rising edges on the two inputs CU (count-up) and CD (count-down) with a pre-set value loaded. If the current count value (CV) reaches zero then output QD is set true. If the current count value reaches PV then the output QU is set true. CU = count up on rising edge input (BOOL) CD = count down on rising edge input (BOOL) R = reset (BOOL) LD = load CV with PV (BOOL) PV = pre-set value (INT) QU = count-up output (BOOL) QD = count-down output (BOOL) CV = current count value (INT)
Pulse timer	The pulse timer is used to produce an output pulse of a given duration. The time duration is determined by the input PT. When IN is set true the output goes high for the specified time PT. IN = start pulse timer input (BOOL) PT = pulse duration time (TIME) Q = output pulse (BOOL) ET = elapsed time (TIME)

TRANSITION Trans 1

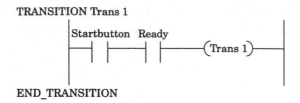

END_TRANSITION

Figure 5.3 Ladder diagram transition expression.

Only one path is selected. Normally
transition conditions are tested from
left to right but precedence can be user
defined

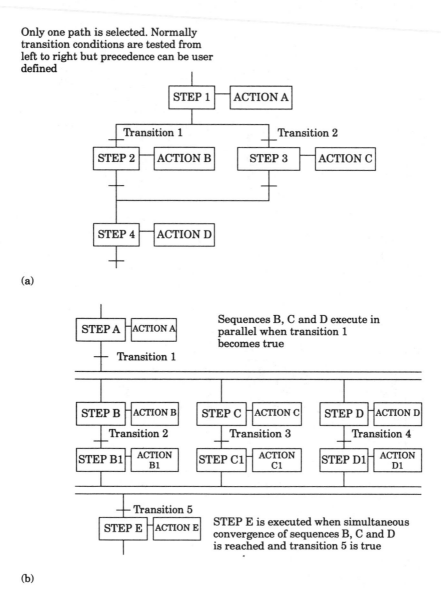

(a)

Sequences B, C and D execute in
parallel when transition 1
becomes true

STEP E is executed when simultaneous
convergence of sequences B, C and D
is reached and transition 5 is true

(b)

Figure 5.4 Sequential function chart structures: (a) use of divergent branches
to enter alternate sequences and (b) parallel sequences.

'stored' since it will continue as the control sequence progresses. A set or stored action is cleared using an R (reset) qualifier. Other action qualifiers are listed in Table 5.2 (e.g. see ACTION/END_ACTION).

Within an SFC it is possible to use a divergent branch to select a path option. This is shown in Fig. 5.4(a). In this example, there are two alternate paths which could be selected according to the transition conditions 1 and 2. Normally the transition conditions are tested from left to right. The flow of control is through the branch for which the transition condition is true. If both transition conditions are true at the same time then the left-hand branch is taken. However, the precedence in which the transition conditions are tested can be user defined.

It is possible to activate sequences in parallel, as shown in Fig. 5.4(b). In this example, two parallel branches are constructed and sequence flow is simultaneous through the two branches. A pair of parallel horizontal lines are used to denote the start and end of the simultaneous sequencing of branches. The branch sequences continue independently. Convergence of the branch sequences must be obtained before they can be combined into one sequence (e.g. transition 5 STEP E). Some branching and parallel sequence examples for Mitsubishi PLCs are given in Appendix 3.

5.6 Translating between languages

Simple ladder rungs involving combinational logic can usually be translated into FBD or ST, as illustrated in Fig. 5.5. The IEC 1131-3 standard specifies the operation of the up-counter, down-counter and up-down counter in terms of an ST language algorithm. Consider the function block representation of the up-counter as shown in Fig. 5.6. The operation of the up-counter function block can be specified using the structured text language algorithm shown overleaf.

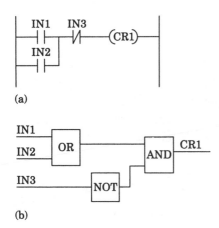

(a)

(b)

CR1: = (IN1 OR IN2) AND (NOT IN3);

(c)

Figure 5.5 Translating a ladder diagram into other IEC languages: (a) ladder diagram, (b) function block diagram and (c) structured text.

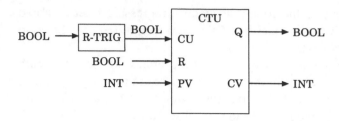

Figure 5.6 Up-counter function block.

```
FUNCTION_BLOCK CTU (*CTU is an abbreviation for
                    CounT Up*)
  VAR_INPUT
    (*data type of inputs*)
    CU : BOOL R_TRIG;        (*Count input on rising
                              edge transition*)
    R : BOOL ;               (*Reset*)
    PV : INT;                (*Pre-set value*)
  END_VAR
  VAR_OUTPUT
    (*data type of outputs*)
    Q : BOOL;                (*Counter output*)
    CV : INT;                (*Current count value*)
  END_VAR
(*Main body of function block*)
IF R THEN CV := 0;
ELSIF CU AND (CV <PV)
  THEN CV:=CV+1;
END_IF;
Q :=(CV>=PV);
END_FUNCTION_BLOCK
```

The up-counter counts the number of rising edges observed at the count-up input CU. The pre-set value (PV) is the maximum count value to be reached. When a new rising edge input is detected the count value (CV) is incremented by one. The output Q is set true when CV is equal to the pre-set value PV. The reset input R can be used to clear Q (e.g. set false) and CV (e.g. set to zero).

References

1. *Programmable Controllers - Part 3: Programming Languages*, International Electrotechnical Commission, IEC 1131-3, 1993 (also British Standard BS EN 61131-3:1993).
2. *Preparation of Function Charts for Control Systems*, International Electrotechnical Commission, IEC 848, 1988.

6
Ladder programming examples

This chapter is basically a programmer's guide to the ladder diagram programming method. As discussed in Chapters 1 and 5, the ladder diagram programming
method has been evolved from conventional electrical circuit relay-based control
methods and many of the concepts are the same.

In essence, a ladder diagram has a left-hand vertical power rail that supplies
notional power through contacts arranged along horizontal rungs. Each contact
represents the state of a Boolean variable. When all contacts in a horizontal rung
are in the true state, power is deemed to flow along the rail and operate a coil on
the right of the rung. In ladder programming terminology contacts can be referred
to as inputs and coils as outputs.

Ladder diagram programming is equivalent to drawing a network containing
switch elements. Modern hand-held programming consoles having liquid crystal
displays allow ladder diagrams to be entered directly. Function blocks such as
timers and counters can be connected into ladder diagram rungs provided that their
inputs and outputs are Boolean variables. Specific details on using function blocks
are given in this chapter.

6.1 Combinational logic

A combinational logic circuit is one in which the function of the output is a direct
and unique consequence of the combination of the input conditions. The basic
logic functions AND, OR and NOT are used in combination logic circuit designs.

A three-input OR gate has an output f, given by

$$f = A + B + C$$

where A, B, C are the input Boolean variables. The ladder diagram is drawn as
shown in Fig. 6.1. In this case, the output coil f is true (1) if A is true (1) or B is
true(1) or C is true(1).

A three-input AND gate has an output f, given by

$$f = ABC$$

This has a ladder diagram as shown in Fig. 6.2.

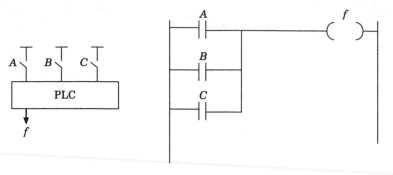

Schematic layout diagram   Ladder diagram

Figure 6.1   Three-input OR function.

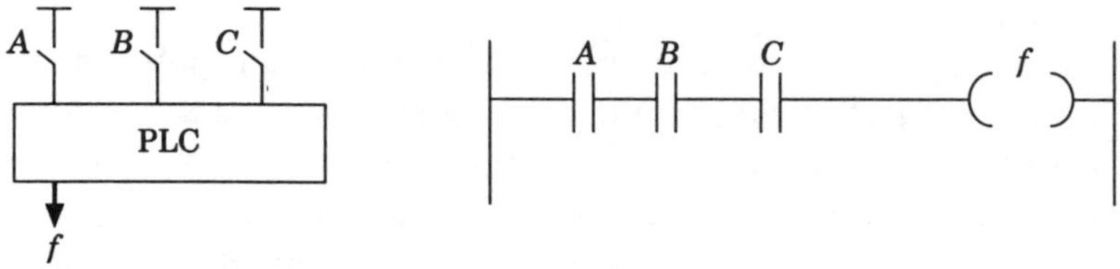

Schematic layout diagram   Ladder diagram

Figure 6.2   Three-input AND function.

In general, any combinational logic output requirement can be drawn as a ladder diagram. For example, the output *f* represented by the function

$f = AB+CD+EF$

is drawn as the ladder diagram shown in Fig. 6.3.

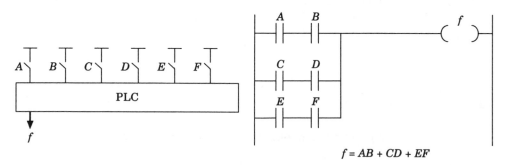

$f = AB + CD + EF$

Schematic layout diagam   Ladder diagram

Figure 6.3   Combinational logic example.

The NOT function can be used to toggle outputs as shown in Fig. 6.4. In this example, the output f_1 equals A (i.e. $f_1=A$) and the output f_2 equals NOT A (i.e. $f_2=\overline{A}$).

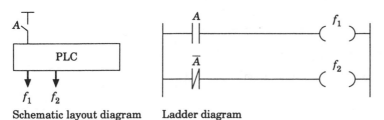

Schematic layout diagram   Ladder diagram

Figure 6.4   Toggling outputs using the NOT function.

The output f of a two-input NAND function is given by

$$f = \overline{AB}$$

The ladder diagram can be drawn by making use of De Morgan's theorem, which states that

$$f = \overline{AB} = \overline{A} + \overline{B}$$

The ladder diagram for the two-input NAND function is shown in Fig. 6.5. It is drawn as NOT A in parallel (i.e. OR) with NOT B.

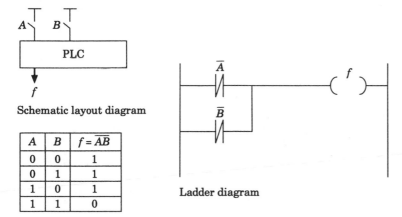

Schematic layout diagram

A	B	$f = \overline{AB}$
0	0	1
0	1	1
1	0	1
1	1	0

Truth table

Ladder diagram

Figure 6.5   NAND ladder diagram.

Similarly, by using De Morgan's theorem, the output f of the two-input NOR function can be written as

$$f = \overline{A + B} = \overline{A}\ \overline{B}$$

The ladder diagram is drawn as NOT A in series with NOT B. This is shown in Fig. 6.6.

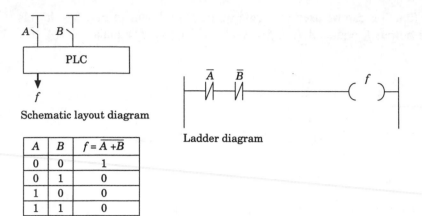

Schematic layout diagram

Ladder diagram

A	B	$f = \bar{A} + \bar{B}$
0	0	1
0	1	0
1	0	0
1	1	0

Truth table

Figure 6.6 NOR ladder diagram.

The exclusive OR function (written as XOR) is a special form of the OR function whose Boolean expression is

$$f = \bar{A}B + A\bar{B}$$

Its ladder diagram can be drawn as shown in Fig. 6.7.

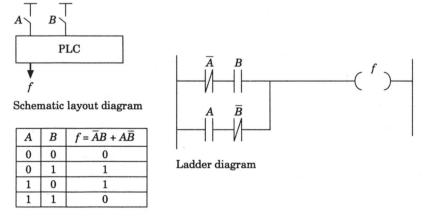

Schematic layout diagram

Ladder diagram

A	B	$f = \bar{A}B + A\bar{B}$
0	0	0
0	1	1
1	0	1
1	1	0

Truth table

Figure 6.7   XOR ladder diagram.

6.2   Latching

A ladder latch circuit allows an output coil to be held (set) and maintained on until a different condition occurs which is used to reset the coil off. An example of a latch circuit is shown in Fig. 6.8.

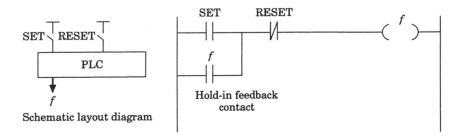

Figure 6.8 Latch circuit.

Assuming all contacts are in their initial states, the output f is set true when the input SET becomes true. Once set the output f is no longer determined by the input SET as it keeps itself maintained on through its feedback contact f. In relay circuit terminology the feedback contact is referred to as the maintaining or hold-in contact. The normally closed contact RESET can be used to clear the latch. The IEC standard 1131-3 defines a set coil symbol and a reset coil symbol which allows the associated variable to be latched on and off (see Fig. 6.9).

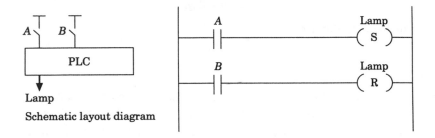

Figure 6.9 Latching using set and reset coils.

The latch is a sequential circuit and can be considered as the 'flip-flop' of ladder logic. The term flip-flop reflects a device that can flip into one state or flop back again to the other on command. The latch (i.e. flip-flop) represents a basic memory device because the output remembers the state of the 'set' input after it has been removed. Sequential devices always incorporate a memory function so that the output can depend on previous as well as present inputs. Counters and timers are other examples of sequential devices.

6.3   Generating a pulse signal

Figure 6.10 shows the characteristics of a positive-going logic pulse. The edge transition from 0 to 1 is defined as the rising or leading edge. The edge transition from 1 to 0 is defined as the falling or trailing edge. The width of the pulse is referred to as the pulse width.

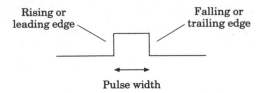

Figure 6.10   Pulse characteristics.

Many PLC systems allow a memory bit to be pulsed on command for a fixed duration of one program execution cycle. A typical circuit is that shown in Fig. 6.11. When the input A is turned on (true) a positive pulse is generated on the memory element MX5 (memory bit 5). This memory element can be used elsewhere in the ladder diagram, for example to reset a counter. The IEC function block symbol for a general purpose pulse timer is shown in Table 5.5 (Chapter 5).

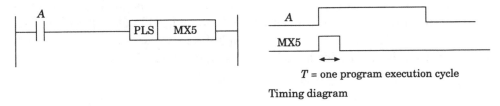

Figure 6.11   Pulsing a memory bit.

6.4   Timers

An example of an on-delay timer is shown in Fig. 6.12. The timer set value (i.e. pre-set time) is 5 seconds (T#5s). When the input A becomes true the timer is activated and when the specified set value has elapsed the output is set true. Consequently, the action of an on-delay timer is to delay setting the output lamp on, as shown in the timing diagram. The timer is reset when the input A goes low.

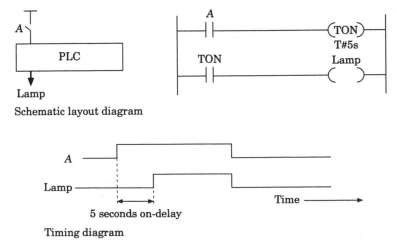

Figure 6.12   On-delay timer.

An example of an off-delay timer is shown in Fig. 6.13. The timer set value (i.e. pre-set time) is 5 seconds (T#5s). When the input *A* goes low the timer is activated and holds the output lamp on for the specified set value. Consequently, the action of an off-delay timer is to delay setting the output lamp off, as shown in the timing diagram.

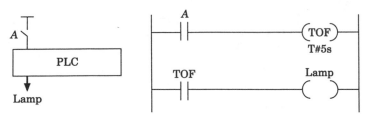

Schematic layout diagram

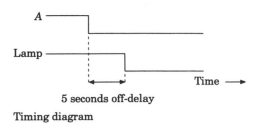

5 seconds off-delay

Timing diagram

Figure 6.13 Off-delay timer.

If the period for which the input is turned on is shorter than the set value (pre-set time) of a timer the output does not change state. The IEC symbols for function block on-delay and off-delay timers are shown in Table 5.5 (Chapter 5). Generally, proprietary PLC systems use a coil representation for timers in ladder diagrams rather than a function block (see Appendices 3 and 4).

6.5 Counters

An example of an up-counter block incorporated in a ladder diagram is shown in Fig. 6.14. A positive pulse applied using the reset contact clears the counter so that Q is false and the pre-set value (PV) is 3. If three positive-going pulses are applied using the count contact the output is set true on the rising edge of the third pulse. Any further pulses applied to the input using the count contact are ignored (i.e. counting stops) until the counter is reset.

A ladder diagram incorporating an up-down counter is shown in Fig. 6.15. This type of counter allows increment (count-up) and decrement (count-down) input signals to be connected. A reset input and pre-set value (PV) must also be provided. A rising edge applied to the count-up input increases the count value (CV) by one. A rising edge applied to the count-down input decreases the count value (CV) by one. When CV=0 the output QD is set true and further changes on

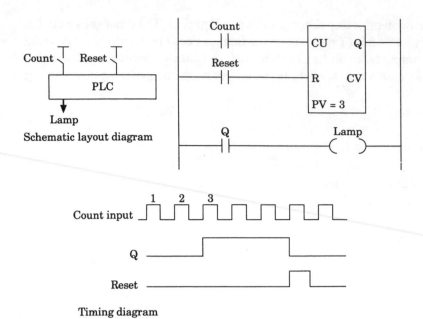

Timing diagram

Figure 6.14 Up-counter.

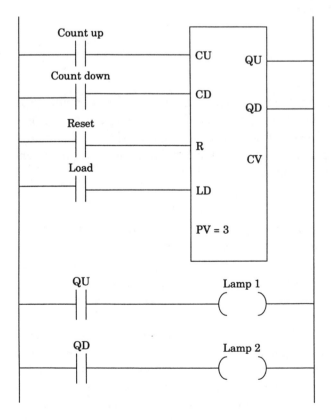

Figure 6.15 Up-down counter.

the count-down input are ignored until a reset signal is applied. When CV=PV the output QU is set true and further changes on the count-up input are ignored until a reset signal is applied.

6.6 Shift register

A shift register is a set of memory elements connected in series in which binary data is stored. Usually the number of memory elements is 4, 8, 16 or 32. There are three inputs, a data input, a shift input and a reset input. The data input is used to transfer a binary value into the first memory location. A pulse applied to the shift input moves all the binary values stored in the register along by one location. The reset input is used to clear all the memory elements so that they store zero.

An example of a 4-bit shift register consisting of memory elements M100, M101, M102 and M103 is shown in Fig. 6.16. The first memory element M100 has the same state as the data input contact on I1. The rising edge of a shift pulse input signal shifts all the data in the register one bit to the left. Assuming the shift register is initially cleared and a logic 1 is entered into M100 a shift pulse moves the logic 1 into M101. Note that the memory elements M100 to M103 are used to drive the output coils Q400 to Q403. Data in the last memory location of the register is lost when a shift is applied.

6.7 Conditional jumps and subroutines

It can be required that part of a ladder program has to be inactive when an input condition occurs. A 'jump' function block can be used to skip ladder rungs on the condition that a contact is made. When a contact is turned on the cyclic scan does not execute ladder rungs between *jump* and *end-jump* labels marked on a ladder diagram.

A subroutine represents a complete ladder program which can be called repeatedly from a main ladder program when required. A subroutine is implemented using a special function coil for calling the routine in the ladder circuit. When the subroutine is called the processor scans the subroutine and returns control to the main program when its actions have been completed. Details of implementation of jumps and subroutine methods are manufacturer specific.

6.8 Arithmetic functions

Most PLCs are able to do simple binary arithmetic operations with data stored in registers using function blocks to add, subtract, multiply and divide binary and BCD numbers. Instructions to set and clear a carry flag are used as binary arithmetic operations make use of a carry in their results (see Appendices 2 and 5).

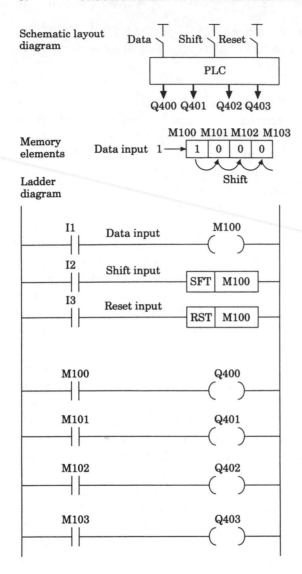

Figure 6.16   A 4-bit shift register.

7
Tutorial examples

The aim of this chapter is to provide some further programming techniques useful for applications. The areas covered include simple minimization of a Boolean equation derived from a truth table, sequencing using timers, a cyclic timer circuit and a cyclic shift register circuit.

7.1 Deriving a ladder diagram from a truth table

In general, any combinational logic process can be described by a truth table which lists the logic values of the outputs for all possible combinations of the input. A truth table with one output and n inputs consists of 2^n rows. For example, a truth table with three inputs has 2^3 or 8 rows.

Given a truth table, an output function can be derived by one of two methods called the *sum-of-products* method and the *product-of-sums* method. The sum-of-products method is applied using the following rules:

1. Pick out all the rows that contain a logic 1 in the output column.
2. For each row that has an output equal to logic 1, logically AND the corresponding inputs together using a negation whenever an input is zero.
3. Take all the row products and OR them together.

For example, consider the truth shown in Table 7.1, where A, B and C are the Boolean inputs to the PLC and f is the output coil to be energized according to the truth table. Applying the sum-of-products method the output is given by

$$f = \overline{A}\overline{B}C + A\overline{B}C + ABC$$

This can be drawn as a ladder diagram as shown in Fig. 7.1.

This expression can be reduced to a simpler form by applying the laws of Boolean algebra to produce a more economical logic circuit. This process of reducing the circuit to a simpler form is called minimization. The steps involved in minimizing this function are described below.

The output can be rewritten as

$$f = \overline{A}\overline{B}C + A\overline{B}C + A\overline{B}C + ABC$$

Table 7.1 Example truth table

A	B	C	f
0	0	0	0
0	0	1	1
0	1	0	0
0	1	1	0
1	0	0	0
1	0	1	1
1	1	0	0
1	1	1	1

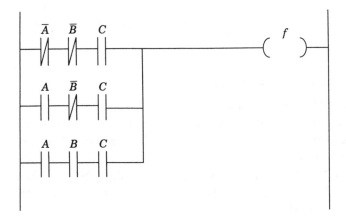

Figure 7.1   Sum of products ladder.

because the idempotent law (1.5) states that

$$A\overline{B}C + A\overline{B}C = A\overline{B}C$$

The reason for expanding the function in this way is because it allows each of the terms to be combined in such a way that they can be reduced. The output can be written as

$$f = \overline{B}C(\overline{A} + A) + AC(\overline{B} + B)$$

Applying the complementarity law (1.13) yields

$$f = \overline{B}C + AC$$

which can be written as

$$f = C(A + \overline{B}) = (A + \overline{B})C$$

This expression for f is called the minimal because it cannot be further reduced. The ladder diagram for the minimal is drawn in Fig. 7.2.

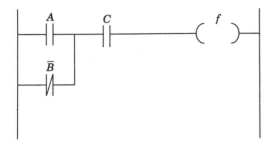

Figure 7.2 Minimal form.

An alternative approach to solving this problem would be to use the product-of-sums method. This is applied using the following rules:

1. Pick out the rows that contain a logic 0 in the output column.
2. For each row that has an output equal to logic 0 logically OR the corresponding inputs together using a negation whenever an input is a logic 1.
3. Take all the row products and AND them together.

7.2 An off-delay timer circuit

Many PLCs provide the programmer with on-delay timers only. An off-delay timer circuit can be created using an on-delay timer as shown in Fig. 7.3. The output Q1 goes high when the input I1 is set high and maintains itself high through the hold-in latch contact. When input I1 is set low the timer resets the latch after the specified pre-set value is reached. Consequently, the output is held for the specified pre-set time after the input is taken low.

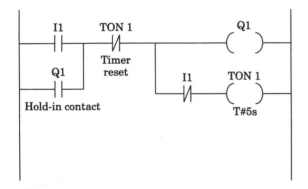

Figure 7.3 Off-delay timer circuit.

7.3 Controlling outputs using timers

Timers can be used to sequence a set of output states. An example circuit is shown in Fig. 7.4. When an input pulse is applied to I1 the outputs Q1, Q2 and Q3 are turned on and off one after the other, as illustrated in the timing diagram. Each output is held on for the specified pre-set time. The PLC scan ensures that the step-to-step sequencing of outputs turning on and off is repeated.

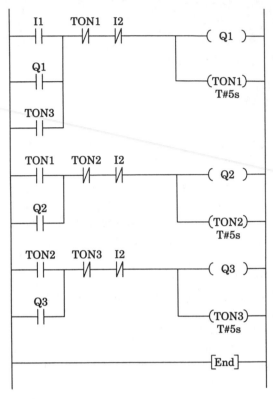

Ladder diagram

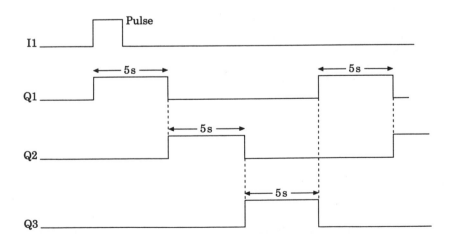

Timing diagram

Figure 7.4 Sequencing using timers.

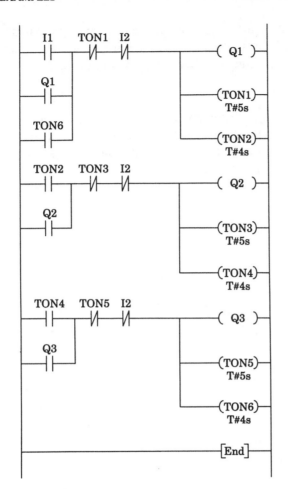

Ladder diagram

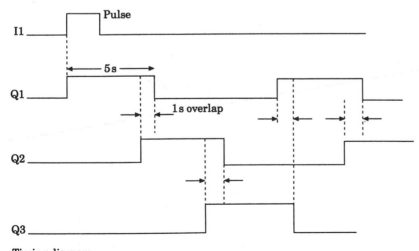

Timing diagram

Figure 7.5   Sequencing with overlap.

Overlap of the output on/off transitions can be achieved by using additional timers to set the next output high before the previous output is taken low. An example ladder circuit and timing diagram are shown in Fig. 7.5.

7.4   Cyclic timer

A ladder circuit that repeatedly turns an output on and off at regular intervals is called a cyclic timer. A cyclic timer circuit is shown in Fig. 7.6. When I1 is set high, the on-delay timers TON1 and TON2 set and reset each other, with the result that Q1 is clocked on and off at 5 second intervals.

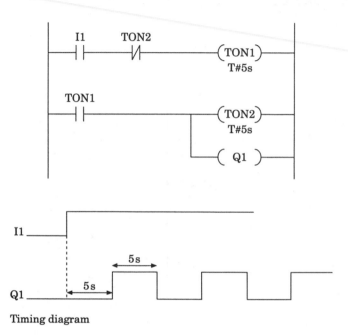

Timing diagram

Figure 7.6   Cyclic timer circuit.

The cyclic timer circuit can be combined with the shift register to time-sequence a set of outputs turning on. An example circuit is shown in Fig. 7.7.

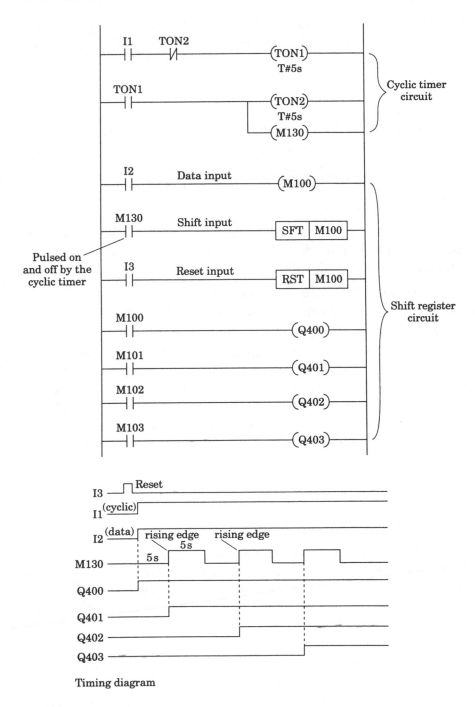

Figure 7.7 Cyclic shift register.

8
Application examples

PLCs are used in a range of industrial applications such as robotics, packaging machinery, pneumatics, conveying and sorting and even greenhouse control. The successful solution of a control problem requires:

- An understanding of the system being controlled
- The assignment of input and output devices to PLC I/O points
- Development of a program

In any control task, the first thing that must be considered is the number and type of input/output points required. This can be assessed by drawing a schematic layout diagram of the system which identifies each input/output device connected to the PLC system. The programming solutions for the application examples of this chapter are expressed as ladder diagrams incorporating function block elements. The exception is the temperature control example in which a structured text code fragment is developed.

8.1   Example 1: control of a pneumatic piston

There are two objectives of this application:

1. To use a PLC to move the piston A shown in Fig. 8.1 out of its cylinder (A^+ direction) when the push-button S1 is pressed and to move the piston into its cylinder (A^- direction) when the push-button S2 is pressed.
2. To continuously move, with the use of a timer, the pneumatic piston in and out of its cylinder at pre-set time intervals.

The control elements for this application are shown in Fig. 8.1. The pneumatic cylinder comprises a piston which is driven by compressed air. The air flow direction to the cylinder is controlled by the solenoid-driven 5/2 (i.e. five port and two air flow directions) pneumatic valve. When the solenoid Y1 is energized the air flow direction is such that the piston moves in the A^+ direction. When the solenoid Y2 is energized the air flow direction is such that the piston moves in the A^- direction.

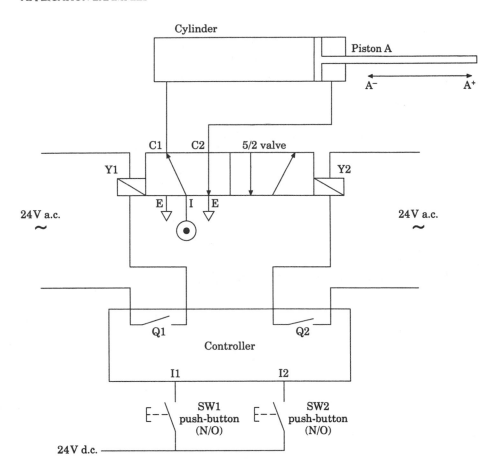

Figure 8.1 Control of a pneumatic piston.

A ladder program solution for the first objective is shown in Fig. 8.2. Two latch circuits are used. When I1 is momentarily taken high the output Q1 is energized and held on via its latch contact. When I2 is momentarily taken high the output Q1 is de-energized (i.e. I2 resets the latch) and Q2 is energized and held on via its latch contact. The Q2 latch circuit is reset when I1 is taken high. Consequently, only one of the solenoids Y1 and Y2 are on at any one time.

A ladder program solution for the second objective is shown in Fig. 8.3. It is based on a cyclic timer circuit discussed in Chapter 7. The two timers TON1 and TON2 set and reset each other at 5 s intervals, with the result that Q1 and Q2 are toggled (i.e. switched so that when Q1 is on Q2 is off and vice versa) every 5 s. Consequently, the piston moves in and out of its cylinder every 5 s.

8.2   Example 2: sequencing three pneumatic pistons

The objective of this application is to use timers to operate three pistons such that the sequence A^+, A^-, B^+, B^-, C^+, C^- is repeated. The three pneumatic pistons shown

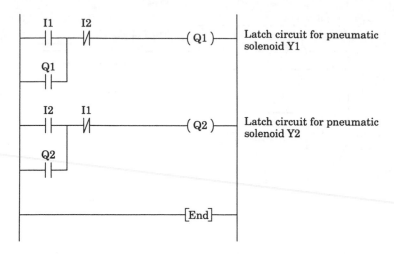

Figure 8.2 Circuit for controlling the pneumatic piston.

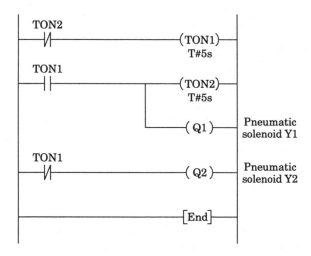

Figure 8.3 Cyclic timer circuit for controlling the piston.

in Fig. 8.4 are controlled using the three 5/2 directional control valves. Each sole-noid has to be held energized for 5 s to allow time for the piston to move. A ladder solution to this problem is shown in Fig. 8.5 on page 92.

8.3   Example 3: counting and packaging

The objective of this application is to continuously divert components down one of two chutes. Figure 8.6 on page 93 illustrates the application, the assignment of PLC input/output points and a ladder program solution. The requirement is for ten components to be directed down chute A and twenty components down chute B for packaging purposes.

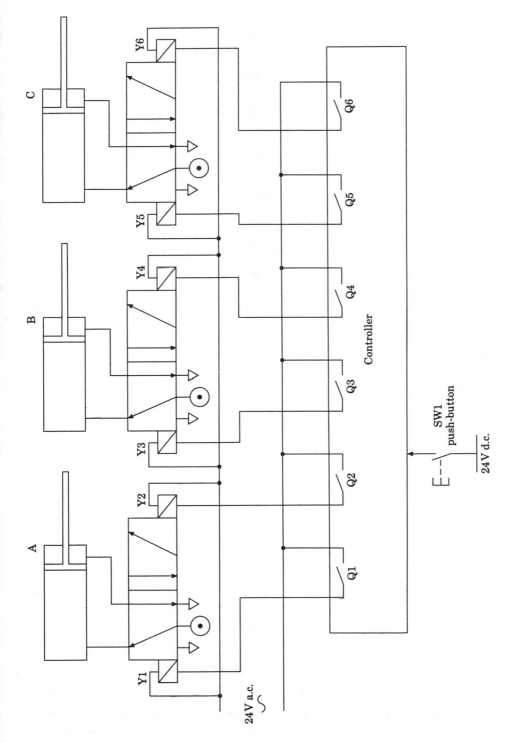

Figure 8.4 Control of three pneumatic pistons.

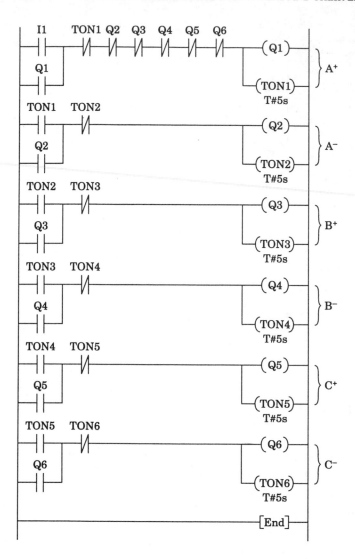

Figure 8.5 Circuit for sequencing the pistons.

A solenoid-driven flap is used to direct components down either chute A or chute B. When the solenoid is energized components are routed down path B. When the solenoid is de-energized the flap (via a spring return mechanism) positions itself so that the components are routed down chute A. The photoelectric switch is used to count components travelling along the conveyor.

A ladder program solution is shown in Fig. 8.6. It uses two counter function blocks. The first counter is used to direct ten components down chute A. The second counter is used to direct twenty components down chute B. A positive pulse on I1 resets the counters and the flap solenoid is de-energized so that components are initially routed down chute A. When the photoelectric switch counts ten components the output Q1 of counter CTD1 energizes the flap solenoid so that

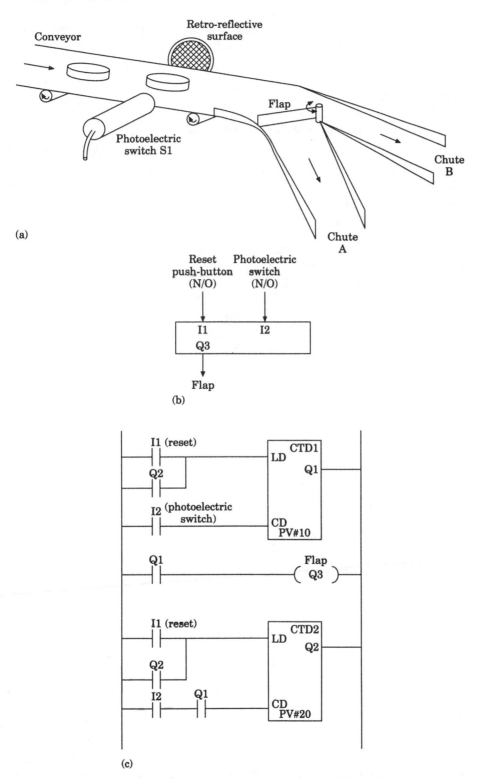

Figure 8.6 A batching application example: (a) application, (b) layout diagram and (c) ladder diagram.

components will be routed down chute B. It also enables count pulses from the photoelectric switch to pass to the count input of the second counter CTD2. When twenty components are counted by CTD2 its output Q2 resets counter CTD1 and itself. When counter CTD1 is reset the flap solenoid is de-energized and components are again routed down chute A.

8.4   Example 4: component detection and sorting

The objective of this application is to detect and sort two geometrically different components travelling along a conveyor. This is a common requirement in packing lines.

The example application is illustrated in Fig. 8.7. Two components labelled A and B travel along a conveyor. Component B is to be deflected by a sort solenoid

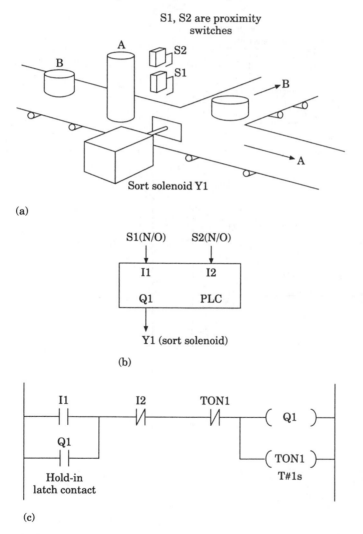

Figure 8.7   Component detection and sorting: (a) application, (b) layout diagram and (c) ladder diagram.

while component A is to be allowed to pass the sort area. Two proximity switches, S1 and S2, are located at different heights to produce a sorting signal. When component A is in the sort area both proximity switches S1 and S2 turn on. When component B is in the sort area only switch S1 is turned on. The sort solenoid is to be energized for 1 s to deflect all components of type B down a different conveyor feed path.

A PLC layout diagram and ladder program solution is shown in Fig. 8.7. Proximity switches S1 and S2 are connected to inputs I1 and I2 respectively. If input I1 (i.e. S1) goes high but not I2 (i.e. S2) a component of type B is registered and the sort solenoid connected to coil Q1 is activated for 1 s. This is achieved using a latch circuit which is reset using the timer TON1. Note that when a component of type A passes the sort area the solenoid is prevented from being activated as the normally closed contact I2 breaks the power flow path to Q1.

8.5   Example 5: pick and place unit

The objective of this application is to show how position switches can be used to control the movement of an axis of a pick and place unit. The pick and place unit shown in Fig. 8.8 has one rotary (θ) and two linear (X and Z) degrees of freedom. The motions are on/off in nature, being actuated by pneumatic pistons. For example, the X motion is in either the X^+ position or the X^- position. The gripper can also be pneumatically operated, being either open or shut. All motions are controlled by actuating appropriate solenoids.

Position switches can be mounted on the unit so that they turn on when a particular motion has occurred. For example, consider two position switches S1 and S2 mounted so that they register the fully extended X^+ position (i.e. S1 is on and S2 is off) and the fully retracted X^- position (i.e. S1 is off and S2 is on). An example ladder program fragment which uses these signals to sequence the X^+ and X^- movements back and forth is shown in Fig. 8.8.

8.6   Example 6: checking for a missing bottle cap

The objective of this application example is to detect and reject bottles emerging from a filling and capping machine that have not been capped. An application sketch, the assignment of input and output points and a ladder program are shown in Fig. 8.9.

Bottle caps can be detected by using two retro-reflective photoelectric switches. One detects the presence of a bottle (i.e. S1) and the other detects the presence of a cap (i.e. S2). The photoelectric switch S1 is mechanically mounted so that it is triggered just before the cap photoelectric switch S2. S1 is connected to input point I1 and S2 is connected to input point I2. Therefore S1 starts timer TON1 while S2 resets TON1. If a cap signal is not detected within 2 s once the timer TON1 has been triggered with a bottle signal the solenoid-driven reject piston is energized by TON1. The reject solenoid is held energized for 1 s by the timer TON2. The timer TON1 is used to overcome any problems relating to operation timing due to the mounting of the sensors.

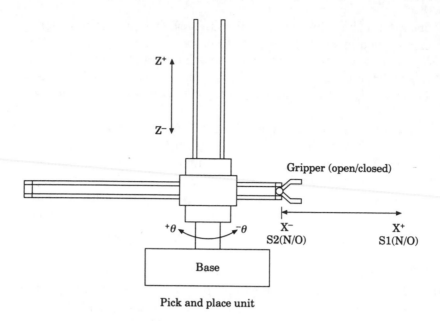

Pick and place unit

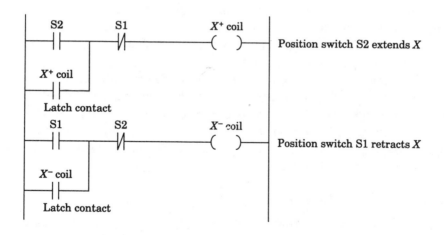

Figure 8.8   Using limit switches to control the X^- and X^+ movements of a pick and place unit.

8.7   Example 7: PID temperature control

Applications such as temperature control require a feedback or closed-loop control system. The basic block diagram of a closed-loop system for temperature control is shown in Fig. 8.10. The temperature sensor is connected to an ADC which generates a binary number according to the temperature measured. The heater is controlled by a DAC. The controller attempts to vary the input to the heater to make the measured temperature equal to the desired set-point value.

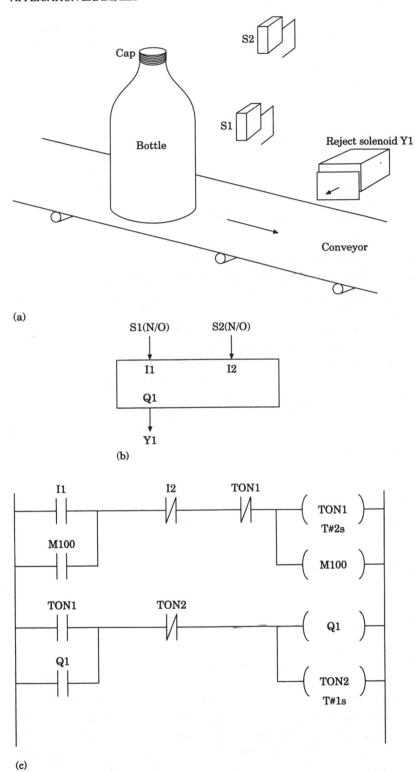

Figure 8.9 Checking for a missing bottle cap: (a) application, (b) layout diagram and (c) ladder diagram.

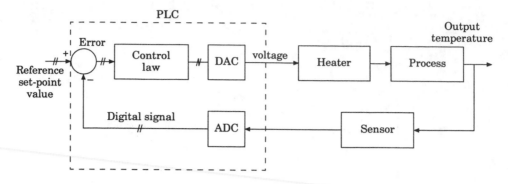

Figure 8.10   Closed-loop temperature control.

The comparator detects the difference between a reference or set-point value and the output of the ADC which represents the measured temperature. This signal is called the error signal. The error signal is fed to the controller and is used to generate an output signal.

Various types of control algorithms are possible which convert the error signal into an output signal for driving the heater. The classical three-term PID (proportional, integral, derivative) control law is widely used. In this case the output is made up of the sum of three terms. The first is proportional to the error, the second is proportional to the integral of the error and the third is proportional to the derivative or rate of change of error. The PID control action equation can be written as

$$x = K_p \left(e + \frac{1}{T_i} \int e \, dt + T_d \frac{de}{dt} \right)$$

where

x = controller output

K_p = proportional gain

e = error

T_i = integral time

T_d = derivative time

Of course, it is possible to simplify the algorithm so that only the proportional (P) or proportional plus integral (PI) terms are implemented.

The important specifications for a closed-loop control system are transient response, steady state error and disturbance rejection. An input step change applied to a system can be used to determine transient response characteristics (e.g. rise time, overshoot, settling time), as shown in Fig. 8.11. The steady state error is the difference between the input and output of the system when the transient response has decayed to zero. Disturbance rejection is the response to an unwanted signal that corrupts the input or output of a process. A PID controller can be tuned to yield minimum overshoot and steady state error.

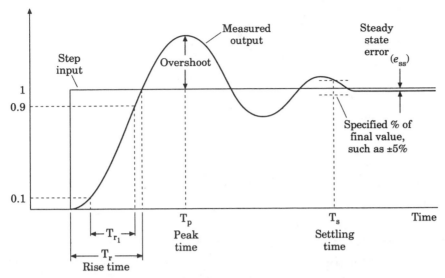

Note: T_r is the rise time, T_{r_1} is the 10–90% rise time.

Figure 8.11 Dynamic response of an under-damped system to a step input.

The Ziegler–Nichols empirical methods for estimating the controller settings of gain, integral time and derivative time can be applied when little is known about the dynamics of the system. There are two approaches that can be used, which are:

● Open-loop step response test
● Closed-loop response test

In the first method, an open-loop step response is performed on the process before the controller is installed. This is achieved by applying a step change to the input and monitoring the output response. By drawing a tangent to the measured output response curve two parameters L and T are obtained, as shown in Fig. 8.12. The controller settings for the P, PI and PID control actions are calculated using the two parameters L, T and the steady state gain K (i.e. the ratio of B/A), as shown in Table 8.1.

Table 8.1 Controller settings for the Ziegler–Nichols open-loop test

Control action	Proportional gain K_p	Integral time T_i	Derivative time T_d
P	$T/(LK)$		
PI	$0.9T/(L\,K)$	$L/0.3$	
PID	$1.2\,T/(L\,K)$	$2L$	$0.5L$

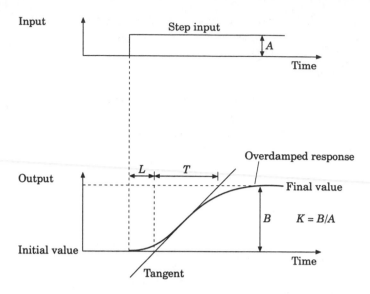

Figure 8.12   Extracting Ziegler–Nichols parameters from an open-loop step response curve of a process.

In the closed-loop method, the controller is installed and set to proportional action only with the integral and derivative terms made inoperative. Starting with a small value, the controller gain K_c is progressively increased in stages until a small step change in the set-point value causes a continuous fixed amplitude oscillation of the output temperature as measured with the sensor. The controller gain K_{osc} and the period of the oscillations T_{osc} are used to calculate the controller settings as shown in Table 8.2.

Table 8.2   Controller settings for the Ziegler–Nichols closed-loop test

Control action	Proportional gain K_p	Integral time T_i	Derivative time T_d
P	$0.5\,K_{osc}$		
PI	$0.45\,K_{osc}$	$T_{osc}/1.2$	
PID	$0.6\,K_{osc}$	$T_{osc}/2$	$T_{osc}/8$

The PID equation is implemented digitally using numerical integration and differentiation using error values obtained at sampling intervals of τ seconds. Assuming that the sampling interval time remains constant, the integral term can be represented as the summation of all the sampled error values multiplied by the sampling interval. Likewise, the derivative term can be represented as the difference between two successive sampled error values divided by the sampling interval (i.e. the slope of the line joining two error samples separated by a sample interval).

A skeletal structured text code fragment which demonstrates the form of a PID algorithm is given below:

```
FUNCTION_BLOCK PID_LOOP

VAR _INPUT
c  : REAL;              (*temperature sensor reading*)
r  : REAL;              (*reference or set-point value*)
e  : REAL : = 0;        (*initialize error to zero*)
Ti : REAL;              (*integral time*)
Td : REAL;              (*derivative time*)
T  : REAL;              (*sampling interval*)
Ki : REAL: = T/Ti;      (*Ki is the sampling interval
                           divided by the integral time*)
Kd : REAL = Td/T;       (*Kd is the derivative time
                           divided by the sample interval*)
K  : REAL               (*proportional gain*)

END_VAR

VAR_OUTPUT
output : REAL;          (*PID output*)
END_VAR

(*PID algorithm*)

REPEAT

e_old := e;             (*old error is made equal to the
                           previous value of error e*)
r := set-point ;        (*read set-point reference
                           value*)
c := temperature;       (*read temperature sensor*)
e := c - r;             (*calculate new error*)
diff := e - e_old;      (*calculate the difference
                           between the new and the previous
                           error)
sum := sum + e;         (*add the new value of e to the
                           accumulated sum*)

output := K * (e + Ki * sum + Kd *diff);
                        (*PID output calculation*)

UNTIL stop_condition;
END_REPEAT
END_FUNCTION_BLOCK
```

8.8 Example 8: PWM speed control of a d.c. motor

PLCs such as the Toshiba Prosec T1 range (see Appendix 5) have a PWM output which can be used for applications such as heater control or motor speed control. For example, the circuit shown in Fig. 8.13 based around a switching power transistor can be used for PWM speed control of a d.c. motor. In this circuit, the motor

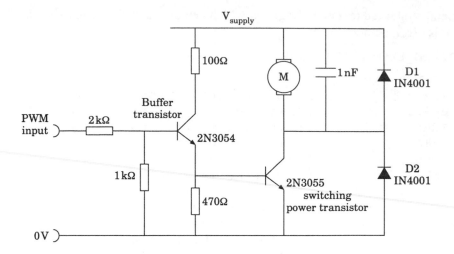

Figure 8.13   Transistor switch circuit for PWM motor speed control.

is fully on when the input is high and is fully off when the input is low. By varying the ratio of the on-time to off-time the speed of the motor can be proportionally controlled. The diodes D1 and D2 in the circuit are used to damp the motor's back e.m.f. and the 1nF capacitor limits unwanted radio-frequency interference.

The average voltage V_{av} applied to the motor is given by

$$V_{av} = D V_s$$

where D is the duty cycle and V_s is the supply voltage. The duty cycle D is defined in terms of the on-time t_{on} and the off-time t_{off}:

$$D = \frac{t_{on}}{t_{on} + t_{off}}$$

The switching frequency f is equal to the reciprocal of the period $(t_{on} + t_{off})$:

$$f = \frac{1}{t_{on} + t_{off}}$$

Therefore the duty cycle can be expressed as

$$D = f t_{on}$$

The duty cycle can be varied in one of two ways. The first method, known as pulse width modulation or PWM, is to keep the frequency f constant as t_{on} is varied. A PWM signal is illustrated in Fig. 8.14. With this signal input the power transistor switches the supply voltage on and off at a fixed frequency with variable mark-space periods so that the average current through the motor load is varied. The second method, known as pulse frequency modulation, is to keep t_{on} constant and vary the frequency.

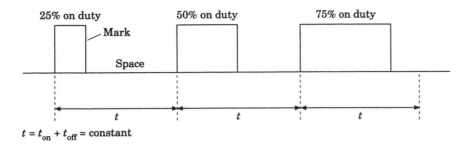

$t = t_{on} + t_{off} = \text{constant}$

Figure 8.14   Pulse width modulation (PWM) signal.

9
Communications

This chapter is concerned with getting two or more pieces of equipment to transfer information. The information transfer may involve a point-to-point link such as a computer to PLC or a network of various types of devices. All communication interfaces are either parallel or serial in nature.

9.1 Parallel interfaces

Parallel interfaces use a group of binary digit data lines (commonly eight) for connecting a PLC with an I/O device. The most common standard is the printer interface based on the Centronics pin assignment (see Fig. 9.1). This is a one-way, high data rate, short-distance connection from a PLC (or computer) to a printer which incorporates handshaking signals for regulating the flow of data. A strobe

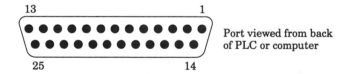

Port viewed from back of PLC or computer

Pin number	Name	Input/output
1	STROBE	Output
2–9	DATA 0–DATA 7	Output
10	ACKNOWLEDGE	Input
11	BUSY	Input
12	PAPER END	Input
13	SELECT	Input
14	AUTO FEED XT	Output
15	ERROR	Input
16	INIT	Output
17	SELECT IN	Output
18–25	SIGNAL GROUND (0 V)	—

Figure 9.1 Parallel port interface connector.

pulse from the PLC or computer transfers parallel data into the printer's internal memory buffer. The acknowledge and busy signals are used for handshaking the data.

9.2 Serial interfaces

In serial data transmission data bits are transmitted one at a time along a single line in what is called a bit stream. Special purpose ICs called UARTs (universal asynchronous receivers transmitters) which convert parallel bus data into serial bit streams are used in serial interface circuits.

Serial data transmission can be synchronous or asynchronous. With synchronous transmission a common clock is used for both the transmitter and receiver. With asynchronous transmission the transmitter and receiver use separate clocks which are synchronized when a character is transmitted. In practice this is done by ensuring that the frequency of the data stream (termed the baud rate) is identical at both the sending and receiving ends. Asynchronous communication is used by most serial data transmission systems.

A serial signal transmitted along a cable as a digital signal (e.g. two voltage levels representing bits 0 and 1) is referred to as a baseband signal. However, the same digital signal could be used to modulate a carrier signal using either an amplitude, frequency or phase-modulating scheme. Systems that employ digital signals to modulate carrier frequencies in the commercial television broadcast band are called broadband systems.

The bandwidth of the conventional public telephone network system is not suited for baseband communication in which the digital signal is put on the cable directly. This is because a digital signal has a frequency response extending from 0 Hz to at least half the bit rate. A telephone line has a poor low-frequency response, being about 50 Hz, and so is not suited for baseband communication. In fact, the effective frequency response of a telephone line extends from 50 Hz to 4 kHz and so if a digital signal is to be transmitted along a telephone line it has to modulate a carrier wave within this frequency range. A device called a modem (an abbreviation for *mo*dulator–*dem*odulator) is used to transfer digital data along a telephone line.

9.3 RS232C

The most commonly used standard for point-to-point serial data transmission is the RS232C standard as defined by the Electronic Industries Association (EIA). The RS232C specification allows for asynchronous communication with data rates up to 19.2 kbaud (k-bits/s), where the communicating devices can be separated by a distance up to 15 m.

RS232 devices are divided into two classes, namely data terminal equipment (DTE) and data communication equipment (DCE). Computers and PLCs are examples of data terminating devices. A modem is an example of a data communicating device. The RS232C signals and pin-out connections are shown in Fig. 9.2.

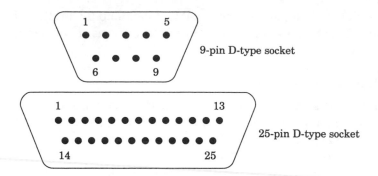

25 pin	9 pin	Label	Direction DTE	Direction DCE	Description
1		Frame GND			Protective ground. It is connected to the metal screening of the cable
2	3	TXD	OUT	IN	Transmit Data line
3	2	RXD	IN	OUT	Receive Data line
4	7	RTS	OUT	IN	Ready To Send. A handshake line which indicates when an RS232C device is ready to send data out
5	8	CTS	IN	OUT	Clear To Send. A handshake line which indicates that an RS232C device is ready to receive data
6	6	DSR	IN	OUT	Data Set Ready. A handshake line to indicate when an RS232C receiving (e.g. modem) device is ready
7	5	SG			Signal ground. The common 0V return line for digital signals travelling along the data link
8	1	DCD	IN	OUT	Data Carrier Detected
20	4	DTR	OUT	IN	Data Terminal Ready. A handshake line used to indicate that an RS232C terminal is ready to start handling data
22	9	RI			Ring Indicator

Figure 9.2   RS232C connectors and pin-out description.

The RS232 standard is intended to link a DTE (e.g. computer) to a DCE (e.g. modem). Consequently, DTE and DCE sockets are connected using a straight-through cable, as shown in Fig. 9.3. Although non-standard, it is also possible to connect two DTEs together (e.g. computer to PLC) using a null modem cable in which crossovers between the transmit and receive data pins and other like pin pairs are inserted (see Fig. 9.3). The term null modem is used because the cable replaces a pair of modems which would otherwise have been used.

A voltage of between +3 and +12 V is used to represent a logic 0 and a voltage of between −3 and −12 V, a logic 1. An RS232C serial bit stream is packaged as a start bit, a data word (usually a 7-bit ASCII character), an optional parity bit and one or two stop bits, as shown in Fig. 9.4. The parity bit provides a way of

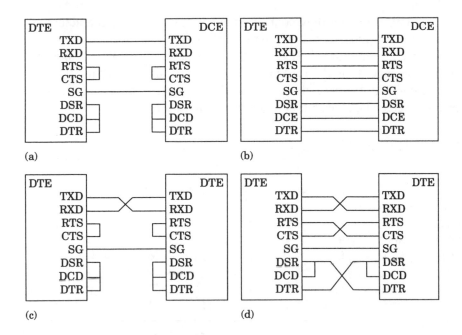

(a)   (b)

(c)   (d)

Figure 9.3 Connections for different RS232C configurations: (a) three-wire DTE to DCE link, (b) straight-through DTE to DCE full connection, (c) three-wire DTE to DTE connection and (d) DTE to DTE full connection.

checking whether data has been corrupted and can be even, odd or disabled. It represents an additional bit added to the data so that a check can be made on the number of ones (or zeros) being either an odd number (odd parity) or an even number (even parity).

Most RS232C communications programs use a memory buffer in which information received through the serial port is stored. A method of flow control is required to stop the transmitting device before the receiver's memory buffer becomes full. Hardware flow control makes use of the handshake signal lines and data is exchanged on a ready (i.e. space available in the memory buffer) and not ready (i.e. buffer is near full) basis. However, it is also possible to use a software

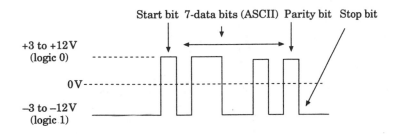

Figure 9.4 RS232C serial data transmission.

hand-shaking method of data flow control whereby stop and start characters are embedded in the transmitted data to regulate its flow.

Data is normally transmitted in the ASCII (American Standard Code for Information Exchange) format. The ASCII character set, also known as ISO 646, is shown in Table 9.1. A 7-bit code is used to differentiate upper and lower case alphanumeric characters, punctuation and control characters.

Table 9.1 ASCII characters

LS	MS	0 000	1 001	2 010	3 011	4 100	5 101	6 110	7 111
0	0000	NUL	DLE	SP	0	@	P		p
1	0001	SOH	DC1	!	1	A	Q	à	q
2	0010	STX	DC2	"	2	B	R	b	r
3	0011	ETX	DC3	#	3	C	S	c	s
4	0100	EOT	DC4	$	4	D	T	d	t
5	0101	ENQ	NAK	%	5	E	U	e	u
6	0110	ACK	SYN	&	6	F	V	f	v
7	0111	BEL	ETB	'	7	G	W	g	w
8	1000	BS	CAN	(	8	H	X	h	x
9	1001	HT	EM	)	9	I	Y	i	y
A	1010	LF	SUB	*	:	J	Z	j	z
B	1011	VT	ESC	+	;	K	[	k	{
C	1100	FF	FS	,	<	L	\	l	|
D	1101	CR	GS	–	=	M	]	m	}
E	1110	SO	RS	.	>	N	↑	n	~
F	1111	SI	US	/	?	O	—	o	DEL

CONTROL CODES

NUL	NULL	SI	SHIFT IN
SOH	START OF HEADER	DLE	DATA LINE ESCAPE
STX	START OF TEXT	DC	DEVICE CONTROL
ETX	END OF TEXT	NAK	NEGATIVE ACKNOWLEDGE
EOT	END OF TRANSMISSION	SYN	SYNCHRONOUS IDLE
ENQ	ENQUIRY	ETB	END OF TRANSMISSION BLOCK
ACK	ACKNOWLEDGE	CAN	CANCEL
BEL	BELL	EM	END IF MEDIUM
BS	BACKSPACE	SUB	SUBSTITUTE
HT	HORIZONTAL TAB	ESC	ESCAPE
LF	LINE FEED	FS	FILE SEPARATOR
VT	VERTICAL TAB	GS	GROUP SEPARATOR
FF	FORM FEED	RS	RECORD SEPARATOR
CR	CARRIAGE RETURN	US	UNIT SEPARATOR
SO	SHIFT OUT	SP	SPACE

The speed at which data is transmitted down a transmission line is often expressed as a baud rate, where one baud equals one bit per second. Although it is possible in theory to use any baud rate, a set of standard transmission speeds are used in practice. These are listed in Table 9.2. The period of each bit is equal to 1/(baud rate). For example, at 4800 baud the bit period is approximately 208 μs.

Table 9.2 Common baud rates used with RS232C

Baud rate
50
75
110
150
300
600
1200
2400
4800
9600
19200
38400
76800

When estimating the useful data transfer rate the additional synchronizing bits used to recover asynchronously transmitted data should be taken into account. For example, a total of 11 bits may be required to transmit a 7-bit data character. In this case a line baud of 1200 represents a useful data rate of 764 bits per second (i.e. 764 = (7/11) x 1200).

9.4 RS422 and RS485

The term multidrop describes a serial communications system in which a number of devices can share a single line. The RS422 and RS485 serial communications standards have been introduced to enhance the RS232C specification and to facilitate a multidrop connection.

The key features of the RS422 standard is that up to ten receivers can be driven by a single transmitter using differential signal lines. A differential scheme measures the difference in voltage between two wires and therefore does not need to reference the signal to a common 0 V point, as in the RS232C case. A noise signal which occurs equally on both lines is cancelled out when the voltage difference is measured. Consequently, a very good noise immunity can be achieved which means that higher data rates or longer distances can be used.

RS485 is a multidrop serial communications standard normally used for a master/slave network installation. An RS485 master/slave network configuration has one master node which is usually a host computer and up to 31 slaves which are usually PLCs. The standard allows for 32 physical tri-state devices to share a single serial line. An inactive device is set to a high impedance state which effectively disconnects it from the shared line. RS485 uses differential drivers and receivers. The relationship between baud rate and transmission line length is logarithmic. At 72 kbaud, line lengths of up to 1200 m can be used. At 1.25 Mbaud, line lengths of up to 18 m can be used.

9.5   Distributed control and networks

A distributed control system (DCS) is one in which several controllers (each with their own processor) are operating simultaneously and co-operatively. Such systems can be divided into tightly coupled multiprocessor systems sharing a common bus back plane (e.g. VME rack) and loosely coupled systems where processor-based controllers are located at the points of control and exchange information via some kind of input/output port. A system of loosely coupled processor-based systems (e.g. computer, PLCs etc.) interconnected together is called a network.

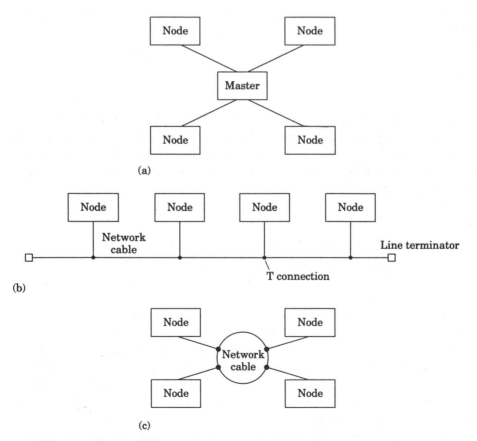

Figure 9.5   Network topologies: (a) star, (b) bus and (c) ring.

To manage an entire plant, PLCs are connected in a network. This can be based on an international standard such as Ethernet or is proprietary. Networks are classified by the topological structure of their interconnections. There are three principal configurations, namely star, bus and ring, as illustrated in Fig. 9.5. A computer, PLC or other system on a network is called a node.

With the star topology, all nodes are connected to a central or master server, forming a star shape. Separate wires are required for each of the nodes attached to the master. The processing capacity of the central server has to be sufficient to handle the communications traffic from all the other nodes attached. Also, the processing capacity and the number of serial port connections of the central server also need to be scaleable as more outer nodes are attached. If this is not possible the star topology is inflexible for future expansion.

With the bus topology, nodes are connected in series and share a common cable. Because a shared cable is used, some method for determining which node has access to the network at any given time must be established. Two access techniques are master/slave and peer-to-peer. On a master/slave network, one node is the master and all other nodes are slaves. The master sends commands out to slave nodes and they respond appropriately. On a peer-to-peer network, any node can initiate a communication to any other node on the network. Ethernet and Token passing systems, discussed below, are examples of peer-to-peer networks.

A network bus system linked by coaxial cable requires terminating resistors at each end of the cable to reduce reflections that are otherwise present on a mismatched line. A ring system is similar to a bus or line system except that the last node is connected to the first node, forming an endless loop. As there are no ends to the network cable terminating resistors are not required. Optical fibre is a useful transmission medium for the factory environment as it has a good bandwidth and is completely immune to electrical interference.

Networks which are local to one building (i.e. the factory) are referred to as local area networks or LANs. In contrast, computer networks used for long-distance communications are called wide area networks, or WANs. A message formatted for transmission over a network is called a packet. A set of hardware and software conventions governing the way data is formatted and transmitted on a network is termed a protocol. Networks with compatible protocols are linked together with devices referred to as routers. Networks using different protocols can be interconnected via devices called gateways.

9.6 The ISO network model

The ISO (International Standards Organization) advocates a reference model consisting of seven communications levels for networks. The layers range from the physical link (lowest layer) to the application layer (highest layer). These communications layers are briefly described below:

1. *Physical link layer*   The first layer defines the physical link in terms of the mechanical connections that tie nodes together in a network and signal levels on the line.

2. *Data link layer*   This layer defines procedures and control protocols for operating the communication lines.
3. *Network layer*   This layer is concerned with switching and routing data packets between nodes.
4. *Transport layer*   This layer is concerned with dividing data into manageable units called packets.
5. *Session layer*   This layer is concerned with starting and organizing an ongoing dialogue (i.e. session) between a user and resource of the network.
6. *Presentation layer*   This layer is concerned with data interpretation.
7. *Application layer*   This layer is concerned with the application program which makes use of the network.

Standards have been developed which cover the two lowest layers (e.g. layers 1 and 2). These include IEEE-802.3 (Ethernet), IEEE 802.4 (token bus) and IEEE 802.5 (token ring). Of these the Ethernet standard is the most widely known. The vast majority of personal computer based LANs use Ethernet adaptors in the form of PC bus interface cards. Ethernet is also becoming the common standard for networking factory floor controllers (e.g. PLCs) with office computers. Siemens, for example, markets an Ethernet module under the proprietary name of Sinec H1 for its midrange PLCs.

9.7   Ethernet

The Ethernet or IEEE 802.3 network standard covers both the physical link and datalink layers of the ISO model. It defines the infrastructure on which the network is built but not the network operating system, which is defined by the higher layers. Consequently, it specifies such things as the type of cable to use (i.e. 50 Ω coaxial cable), the maximum length of cable (i.e. 500 m before a repeater is required), the connecting method (i.e. a BNC T-piece connector) and the data rate (i.e. 10 Mbaud).

The Ethernet standard uses baseband signalling (i.e. bit 1 is represented by one voltage level and bit 0 by another voltage level) and a technique called CSMA/CD (an acronym for carrier sense multiple access with collision detection) to control the flow of information on the network. Using CSMA/CD a device (i.e. computer or PLC with a network adaptor) first checks to see if the cable is free for transmission before it transmits data. If the cable is free for transmission (i.e. no carrier sensed) a packet of data is placed on the network. Because it is possible for two or more devices to start transmitting at the same time it is necessary for each device to monitor the cable for a data collision. If a data collision is detected, the transmitting devices stop and have time-out for a random period before trying to gain access to the network again.

Consequently, devices sense data on the network and wait until it is clear. This means that only one device transmits at any one time. Each data packet transmitted on the network contains the destination address of the node to which it wishes to pass the message. All nodes receive the data packet and compare the destination address with their own address; if a match is found the message is collected. Otherwise it is discarded.

9.8   Token bus and ring

With the token bus and ring systems a device has to wait until it receives a special signal called a 'token' before it gains access to the network. Only the device having the token can communicate over the network. The data on a token bus system is always carrier modulated. The token ring is a baseband system (i.e. digital data transmitted is represented as discrete levels on the cable).

9.9   PLC proprietary networks

Most PLC networks are proprietary, being specific to a particular manufacturer. The PLCs are located at the points of control and control tasks are run simultaneously on separate PLCs. Proprietary networks have their own protocol, which is generally not compatible with that of another manufacturer. Some proprietary networks are listed in Table 9.3 (see also Appendix 6).

Table 9.3 Proprietary networks

Manufacturer	*Network*
Mitsubishi	Melesecnet (II0
Omron	Sysmac
Toshiba	Toshline-F10
	Toshline-30
Siemens	Sinec H1 (Ethernet) Sinec L2 (Profibus) Sinec S1 (ASI bus)
Allen Bradley	Data Highway (plus)
GEC Industrial Controls	Starnet Coronet
Gould Electronics	Modbus

The Mitsubishi Melsecnet (II) is a typical example of a proprietary network. It links two separate network loops (tiers) of PLCs controlled by a personal computer (PC). Each network loop can have 64 PLC units attached. The network cable has forward and reverse dual redundant loops so that if one loop fails the other takes over immediately. The network continues to operate if a PLC fails or is taken off-line for maintenance purposes. The network can be used to collect production data at a central point on the PC.

The response time of a network (also called the access time) is the time taken for two nodes to communicate. It effectively gives a measure of how fast data can be transferred from one PLC to another. The response time increases as the number of active PLC nodes on the network increases. Emergency stop signals should never be sent on a network but should be hardwired.

9.10   The IEC 1131-5 standard

Currently under development is an IEC 1131-5 standard covering PLC network communications. This will sit on top of the ISO application layer and will propose a set of standard communications function blocks to exchange information and data between IEC compliant PLCs. The programming aspects of these function blocks will relate to the PLC programming standard IEC 1131-3 and will provide mechanisms to read and write to variables declared as type 'access'.

9.11   Sensor/actuator bus technology

A variety of sensor/actuator bus systems exist which are used to connect process sensors and actuators to high-level systems such as PLCs. The actuator–sensor interface (ASI) is a leading proprietary product developed by eleven control component manufacturers. Intelligent sensor modules (e.g. a proximity switch incorporating ASI bus interface circuitry) connect to the network bus in the form of a flat cable. All elements within the system communicate on the master–slave principle. Masters are currently available in the form of I/O interface modules for computers using RS232C and Siemens PLCs. Siemens market the ASI bus under the proprietary name of Sinec S1. Interbus-s is another example of a sensor/actuator bus system. Apart from reduced wiring costs the advantage of sensor/actuator networks is the flexibility of connecting an I/O device at any position in the installation.

9.12   SCADA

SCADA (supervisory control and data acquisition) software runs on a master computer and provides protocols to communicate with a variety of remote devices such as PLCs, data loggers and intelligent loop controllers. Such software is valuable in any manufacturing process in which there is a need to collect and record plant data (liquid levels, pH levels, the status of motors, the presence of components, alarms, set-points, etc.) so that overall performance and historical trends can be monitored from a central control site. MRP (materials requirements planning) and SPC (statistical process control) require the monitoring and data acquisition of plant data which SCADA packages provide.

A graphical user interface (GUI) provides an interactive visual display of the industrial process as the SCADA system collects and processes information in real-time. SCADA systems can also allow supervisory control over operating data, e.g. offering simple set-point trim facilities. The PLCs in such a system perform the main control functions and gather data which is communicated (via the proprietary network link) to the site SCADA system. The challenge in many plants is that of obtaining information from separate processes in the plant which are controlled by different PLCs and other control systems. The key feature of SCADA systems is that they allow a wide variety of information to be routed along various types of proprietary PLC networks and from other sources.

Appendix 1
Glossary of terms and abbreviations

Absolute encoder   A shaft or linear encoder that generates a unique number for each resolvable position

A.C. or a.c.   Alternating current

Access time   The time taken for data to be read from memory

Accumulator   A CPU register used to hold the results of arithmetic and logic operations

ADC   Analogue to digital converter. A device that converts an analogue signal into a digital value

Address   The numerical assignment of a particular memory location

ALU   Arithmetic logic unit. An arrangement of logic circuits within the CPU that carries out all arithmetic and logic operations

ANSI   American National Standards Institute

ASCII   American Standard Code for Information Interchange

Bandwidth   Information carrying capacity of a communications channel expressed in bits/s. Also defined as the range of frequencies at which the magnitude response of a system is -3 dB below the magnitude response at zero frequency

Baud   A unit of transmission speed in serial communications equal to the number of signal events per second.

BCD   Binary coded decimal

Bit   Binary digit

Boolean data   Data represented as a single bit

Boot   To load an operating system or other program into a newly reset microprocessor system

bps   Bits per second

BS   British Standard

BSI   British Standards Institute

Buffer   A block of memory used for temporary storage

Bus   A set of conductors used for communicating signals

Byte   A group of eight bits

CAD   Computer aided design

CAM   Computer aided manufacture

CIM   Computer integrated manufacture

Clock   A periodic signal used for synchronization

Closed-loop control   A system in which the output signal is measured and fed back to the input point. It is then compared with the input signal to generate a difference signal called the error

CNC   Computer numerical control

Contact bounce   The problem relating to mechanical switches which produce a noisy signal when switched

Compiler   A program to translate a high-level language into machine code

Counter   A function block that gives an output when a set number of pulses have been applied to the input

CPU   Central processing unit. A CPU is a microprocessor and has an internal structure which consists of an arithmetic logic unit (ALU), a program counter and an instruction decoder

CSMA/CD   Carrier sense multiple access with collision detection

Current sinking   The action of receiving current

Current sourcing   The action of supplying current

DAC   Digital to analogue converter

D.C. or d.c.   Direct current

DCE   Data communication equipment (e.g. modem)

Debug   To search for and eliminate mistakes in a program

DTE   Data terminal equipment (e.g. computer or PLC)

Duplex   (see full duplex)

Earth   Electrical safety conductor

EEPROM   Electrically erasable programmable read only memory

EIA   Electronics Industries Association

EPROM   Electrically programmable read only memory

Fail-safe shutdown   The facility of a PLC system to have its outputs take pre-defined states within a specified delay when a power supply voltage drop is detected or an internal failure occurs

FET   Field effect transistor

Fieldbus   A system for the interconnection of distributed devices such as temperature controllers developed by the Instrument Society of America (ISA)

FIFO   First in first out. Describes a data storage system where the first piece of data stored is the first piece to be processed

Flag   A single bit variable used to indicate that some condition has occurred

Flag register   A CPU register which contains single-bit flags such as those to indicate sign, zero and carry

Flash ADC   High-performance ADC which generates an n-bit digital number by performing 2^n comparisons in parallel

Floating-point number   A digital approximation to a real number. A microprocessor stores a floating-point number as a fractional part called the mantissa and an exponent. For example, the number 12.3 is stored as 0.123×10^2 where the mantissa is 0.123 and the exponent is 2

FLOPS   Floating-point operations per second. A measure of a microprocessor's ability to perform calculations with floating-point numbers

Full duplex   A full duplex data link allows the transmission of data simultaneously in both directions

Gateway   A device that connects and allows messages to be communicated between two or more different network systems

Gray code   A binary code in which consecutive codes differ in only a single bit position

Half duplex   A half duplex data link allows the transmission of data in either direction but on an alternate basis. This means that when one device is transmitting the other is listening and vice versa

Hexadecimal   Base 16 number system

Host   The primary computer in a computer to multiple PLCs (i.e. targets) installation

Hz   Hertz. The unit of frequency equal to one cycle per second

IC   Integrated circuit (silicon chip)

IEC   International Electrotechnical Commission (sister organization of the ISO). IEC standards relating to PLCs are:

    IEC 1131-1   General Information (also BS EN 61131-1, 1994)

    IEC 1131-2   Equipment requirements and tests (also BS EN 61131-2, 1995)

    IEC 1131-3   Programming languages (also BS EN 61131-3, 1993)

    IEC 1131-4   User guidelines for selection, installation and maintenance (under development)

    IEC 1131-5   Messaging service specification (under development)

IEE   Institute of Electrical Engineers

IEEE   Institute of Electrical and Electronic Engineers

Image memory   The part of memory where I/O status information is held

Incremental encoder   A shaft or linear encoder which produces a pulse for each resolvable change in position

Interrupt   An externally generated control signal which causes the CPU to temporarily suspend main program execution and transfer control to a special routine to service the interrupt signal

I/O   Input/output

ISO   International Standards Organization

K   Kilo (one thousand). When referring to memory capacity it is equal to 2^{10} or 1024

Ladder diagram   A programming language in which a circuit, consisting of contacts, coils and other elements, such as function blocks, and bounded between left and (optionally) right vertical lines representing notional power rails, is drawn
LIFO   Last in first out. Describes a data storage structure (e.g. a stack) in which the last piece of data stored is the first to be processed
LSB   Least significant bit

M   Mega (one million). When referring to memory capacity it is equal to 2^{12} or 1 048 576
Machine code   Binary numbers that represent CPU instructions
Man-machine interface   Peripheral equipment intended as an operator interface
MTBF   Mean time between failure
MIL   Military standard
MIPS   Million instructions per second. A measure of a microprocessor's ability to process instruction codes
MSB   Most significant bit
MTBF   Mean time between failure
Multidrop   Describes a serial communications system in which a number of devices may share a single line (see tri-state)

N/C   Normally closed
N/O   Normally open

Octal   Base 8 number system
Off-delay timer   A function block that holds its output high for a specified duration when its input changes from high (bit 1) to low (bit 0)
On-delay timer   A function block that delays setting its output high when its input changes from low (bit 0) to high (bit 1)
Operating System   Software that manages microprocessor resources and controls the execution of user programs

PID function   A function block that provides proportional, derivative and integral control actions for a closed-loop feedback system. A PID algorithm can be tuned to minimize overshoot and steady state error
PMC   Programmable motion controller. A microprocessor-based control system dedicated to single- or multi-axis motion control and often incorporating digital I/O
Program counter   A CPU register containing the address of the next instruction to be executed
Programming languages   IEC-defined programming languages are function block diagram (FBD), instruction list (IL), ladder diagram (LD), structured text (ST) and sequential function chart (SFC)

Protocol A set of rules for communication

PWM Pulse width modulation

RAM Random access memory

Real number A number with a decimal point

Register A byte (8 bits), word (16 bits) or long word (32 bits) of memory that is part of a microprocessor as opposed to general purpose memory

ROM Read only memory

RR Result register, also referred to as the accumulator

RS232C Point-to-point serial data communications standard. The maximum data transmission speed is 19.2 kbaud (kbits/s) at a distance of 15 m

RS422 Serial data communications standard with multidrop capability using 5 V differentially driven signals. Data transmission speeds up to 10 Mbits/s are possible

RS485 Multidrop serial communications standard which can be used for a master/slave network configuration. RS485 uses differential drivers and receivers. The relationship between baud rate and transmission line length is logarithmic. At 72 kbaud line lengths of up to 1200 m can be used. At 1.25 Mbaud line lengths up to 18 m can be used

Run The run mode executes the user program stored in memory

Scan PLC program execution loop to continuously read input values and set outputs according to the program requirements

Simplex A simplex data link system allows the transmission of data in one direction only

SPDT Single-pole double-throw switch

Stack An area of memory for temporarily storing data that has a LIFO structure

State The condition of a Boolean variable

Subroutine A program fragment (within a main program) that can be called by the main program

Successive approximation ADC A converter that makes a series of successive approximations to derive the digital value

Tachogenerator A permanent magnet d.c. motor operated as a voltage generator. Its purpose is to generate a d.c. voltage proportional to the velocity (speed) of the drive shaft

Throughput The number of events that can be processed in a given period of time

Tri-state device The output lines of digital devices sharing a bus can take three possible states, namely logic 0 (low), logic 1 (high) and output disconnected (i.e. a tri-state or high impedance state)

TTL Transistor-transistor logic. TTL devices require a single power supply of +5 V d.c. TTL logic levels are:

Logic 0 = 0–0.8 V
Logic 1 = 2.4–5 V

Two's complement   A system for representing negative numbers in binary notation

Watchdog timer   A timer that independently monitors the integrity of a computer system and generates an interrupt signal if not periodically reset by the system. Watchdog timers are used to monitor the internal hardware functions, and/or application program functions, and/or operating system functions

0 V   The zero voltage d.c. return line

Appendix 2
Number systems

The binary, octal and hexadecimal number systems are widely used in the direct referencing of I/O points and digital data. A brief overview of number systems and some examples of converting from one number system to another are given below.

A2.1 Binary numbers

A binary number is expressed by sequences of the digits 0 and 1. As two digits are used it is said to have a base or radix of 2. Each *binary* digit is called a *bit*. This number system is the basis of all digital computing since these two digits can be represented as the true and false levels in a logic circuit. The actual contents of memory consists of binary numbers. These are usually organized in groups of 4 bits (referred to as a nibble), 8 bits (referred to as a byte),16 bits (referred to as a word) and 32 bits (referred to as a long word).

Binary to decimal conversion involves writing down the successive powers of two for each 1 or 0 in the binary number. For example, the binary number 10110101 is converted to 181 denary as shown below:

$$10110101 = (1 \times 2^7) + (0 \times 2^6) + (1 \times 2^5) + (1 \times 2^4) + (0 \times 2^3) + (1 \times 2^2) + (0 \times 2^1) + (1 \times 2^0)$$

$$= (1 \times 128) + (0 \times 64) + (1 \times 32) + (1 \times 16) + (0 \times 8) + (1 \times 4) + (0 \times 2) + (1 \times 1)$$

$$= 181 \text{ denary}$$

When it is necessary to be explicit, the base number is written as a subscript. For example,

$$0110_2 = 6_{10}$$

The binary number 0110 is equal to the denary (base 10) number 6.

Unsigned binary notation assumes all numbers are positive. Binary arithmetic uses the same rules as for denary arithmetic. For example, consider the binary addition:

```
    0000 1111
 +  0000 1111
 _____
    0001 1110
 _____
```

To obtain this result each column of numbers is added up, working from right to left and starting with the units column. Binary numbers in the columns are added such that:

(1) $0 + 0 = 0$ (i.e. addition yields zero)
(2) $1 + 0 = 1$ (i.e. addition yields a single unit)
(3) $1 + 1 = 1\,0$ (i.e. addition yields 2_{10})
(4) $1 + 1 + 1 = 1\,1$ (i.e. addition yields 3_{10})

If the total is 1 or 0, as in cases (1) and (2), this is marked down in the appropriate column. If the result is 10_2, as in case (3), a 0 is marked down in the appropriate column and the 1 is carried forward to the next column. With a digit 1 carried forward it is possible for case (4) to occur. In this event a 1 is marked down in the appropriate column and a 1 is carried forward to the next column.

In unsigned binary arithmetic the result of an operation can require that a carry flag be set. For example, adding the following two 8-bit numbers together produces an answer which is 9 bits long. The extra ninth bit sets the carry flag of a microprocessor's status register:

```
              1111 0000
         +    01010101
         _____
 carry 1      0100.0101
         _____
```

The one's complement of a binary number is obtained by inverting each bit. It represents the logical NOT operation. For example, the one's complement of the binary number 1111 0000 is 0000 1111.

The two's complement system is the most common number system used to represent signed binary numbers in microcomputers. To obtain the two's complement of a binary number each bit is inverted and one is added to the result. The two's complement of a binary number is calculated as follows:

00000101 (binary representation of $+5_{10}$)

11111010 (one's complement)
_____
        +1 (add 1)
_____
11111011 (two's complement representation of -5_{10})
_____

The furthest digit to the left can be used as a sign digit to indicate whether the number is positive or negative. The sign bit is 0 for positive numbers and 1 for negative numbers.

The two's complement system yields the correct sign and result for an arithmetic operation. For example, adding $+5_{10}$ and -5_{10} results in zero, as shown below:

0000 0101 (binary notation for 5_{10})
<u>1111 1011</u> (two's complement representing -5_{10})

<u>0000 0000</u>

If the result of a binary addition carries forward into the sign bit an overflow flag is set in a microprocessor's status (flag) register.

A2.2   Octal numbers

The octal (base 8) number system uses a set of eight distinct digits, 0 to 7. Octal numbers are organized in ascending powers of 8. For example, the octal number $321_8 = 3 \times 8^2 + 2 \times 8^1 + 1 \times 8^0 = 209_{10}$.

Converting a binary number to octal requires it to be split up into 3-bit groups. This is because the octal numbers 0 to 7 are represented by the 3-bit numbers ranging from 000 to 111. For example, the binary number 11111101 is converted to the octal number 375_8 as shown below:

11111101 = 011 111 101 = 375_8

A2.3   Hexadecimal numbers

The hexadecimal number system (base 16) uses a set of sixteen digits (0, 1, 2, 3, 4, 5, 6, 7, 8, 9, A, B, C, D, E, F). The letters A to F represent the numbers 10 to 15 respectively. Hexadecimal numbers are organized in ascending powers of 16. For example, the hexadecimal number $321_{16} = 3 \times 16^2 + 2 \times 16^1 + 1 \times 16^0 = 801_{10}$.

Converting a binary number to hexadecimal requires it to be split up into 4-bit groups (i.e. nibbles). This is because the hexadecimal numbers 0 to F represent 4-bit binary numbers ranging from 0000 to 1111. For example, the binary number 10101100 is represented as the hexadecimal number AC_{16}, as shown below:

1010 1100 = AC_{16}

A2.4   Binary coded decimal (BCD)

The binary coded decimal (BCD) system is a numerical representation in which each decimal digit is represented by a group of 4 bits. The decimal digits 0 to 9 are represented by the binary numbers 0000 to 1001. For example, in binary coded decimal the number 954_{10} is represented as 1001 0101 0100.

Appendix 3
Mitsubishi F1 range PLCs

A3.1   Main system features

The Mitsubishi F1 range of programmable logic controllers (see Figs. A3.1 and A3.2) is designed for use in small- to medium-scale I/O applications. The main system features are:

- Base units up to 60 I/O point capacity
- Extension unit can be combined with a base unit to increase I/O capacity to 120
- LED indication of I/O status
- Simulator switches for inputs

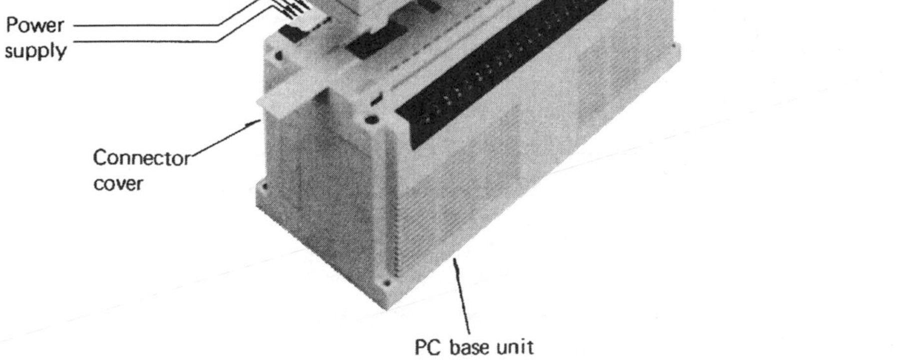

PROGRAMMING PANEL AND BASE UNIT

Figure A3.1   Mitsubishi F1 PLC. *Source*: Mitsubishi Programmable Controller MELSEC F, F1, F2, Series Instruction Manual, Mitsubishi, October 1986.

INSTRUCTIONS AND EXECUTION TIME

Instructions	Designation	Object factors	Execution time[1] ON	Execution time[1] OFF	General functions	
LD	Load	X, Y, M, T, C, S	5.4		Start of logical operation	Normally open contact
LDI	Load Inverse		5.4			Normally close contact
AND	AND		4.2		Logical product (series contact)	Normally open contact
ANI	AND Inverse		4.2			Normally close contact
OR	OR		4.2		Logical sum (parallel contact)	Normally open contact
ORI	OR Inverse		4.2			Normally close contact
ORB	OR Block	None	3.6		Parallel connection of circuit block	
ANB	AND Block				Series connection of circuit block	
OUT	OUT	Y	34.5	34.5	Coil drive instruction	
		M	31.5	31.5		
		S	36.3	48.8		
		T-K	108	142		
		C-K	120	72		
		F671 ~ F675 - K	126	58.9		
PLS	Pulse	M100 ~ M377	49.4	47.0	Rising pulse generating instruction	
SFT	Shift	M100, 120, 140, 160, M200, 220, 240, 260, M300, 320, 340, 360	70.2	50.0	Shift register 1-bit shift instruction	
RST	Reset		63.7	51.8	Reset instruction for shift register, counter	
		C (excluding C661)	44.6	41.7		
S	Set	Y	35.7	29.8	Operation holding coil drive instruction[4]	
		M200 ~ M377	32.7	26.2		
		S	44.6	38.1		
R	Reset	Y	38.1	28.0	Operation holding reset coil drive instruction	
		M200 ~ M377	35.1	25.0		
		S	50.6	32.7		
MC	Master Control	M100 ~ M177	23.8		Common series contact	
MCR	Master Control Reset		3.0		Reset of common series contact	
CJP	Conditional Jump	700 ~ 777	55.4	28.0	Conditional jump	
EJP	End of Jump		0		Designation of conditional jump destination	
NOP	Nop	None	0		None-processing	
END	End		1101[2]		Program end	
STL	Step ladder	S600 ~ S647	14.3 + 69n[3]		Start of step ladder	
RET	Return		14.3		End of step ladder	

[1] It is estimated that one execution cycle time is K times of total execution time calculated from step 0 to END.

$$K = 1.2 + \frac{0.15}{①} + \frac{(0.16}{②} + \frac{0.02}{③} + \frac{0.04)}{④} \frac{f}{⑤}$$

(1) In case T650 T657 are used
(2) In case high-speed counter is used
(3) In case F670 K118 is turned on
(4) In case F670 K121 is turned on
(5) Input frequency of high-speed counter ($f = 1$ for 1kHz)

[2] Input/output processing time is included.
[3] 'n' shows the number of longitudinal connection (parallel joining) for STL instruction.
[4] $51.2 + 31.5n$ for ON and 36.9 for OFF in the STL circuit block.
  n ... Number of longitudinal connection (number of parallel joining) for STL instruction.

Figure A3.2   Basic programming instructions used by Mitsubishi F series *(continued)*.

LIST OF ELEMENT NUMBERS

	00 ~ 13	14 ~ 27	30 ~ 37	40 ~ 47	50 ~ 57	60 ~ 67	70 ~ 77
000's	X: 12 points	X: 12 points	Y: 8 points	Y: 8 points	T: 8 points 0.1 ~ 999s	C: 8 points 1 ~ 999	SPM: 6 points
100's	M: 64 points						
200's	M: 64 points						
300's	M: 64 points						
400's	X: 12 points	X: 12 points	Y: 8 points	Y: 8 points	T: 8 points 0.1 ~ 999 s	C: 8 points 1 ~ 999	SPM: 4 points
500's	X: 12 points	X: 12 points	Y: 8 points	Y: 8 points	T: 8 points 0.1 ~ 999 s	C: 8 points 1 ~ 999	SPM: 6 points
600's	S: 40 points				T: 8 points 0.01 ~ 99.9's	C C: 6 points 1 ~ 999	F: 6 points
700's	CJP/EJP 64 points						

C660/C661
0 ~ 999999 reversible one point

┆┄┄┄┄┆ Extension unit   ☐ Battery back up

X : Input relay   Y : Output relay   M : Aux. relay   SPM : Special aux. relay
T : Timer   C : Counter   S : State   F : Coil for applied instruction

Input/output relay Nos. (basic unit)

Basic unit	Input relay Nos.	Output relay Nos.	Extension connector
F₁ – 12M	400 ~ 405 6p	430 ~ 435 6p	400
F₁ – 20M	400 ~ 412 12p	430 ~ 437 8p	400
F₁ – 30M	400 ~ 413 12p 500 ~ 503 4p	430 ~ 437 8p 530 ~ 535 6P	400
F₁ – 40M	400 ~ 413 12p 500 ~ 513 12p	430 ~ 437 8p 530 ~ 537 8p	400 500
F₁ – 60M	000 ~ 013 12p 400 ~ 413 12p 500 ~ 513 12p	030 ~ 037 8p 430 ~ 437 8p 530 ~ 537 8p	000 400 500

p: points

The value in ☐ of extension unit will be "0", "4" or "5", depending upon extension connector Nos. 000, 400 or 500.

Input/output relay Nos. (extension unit)

Extension unit	Input relay Nos.	Output relay Nos.
F – 4T	☐ 20 ~ ☐ 23 4p	☐ 40 ~ ☐ 43 4p
F₂ – 8EY	—	☐ 40 ~ ☐ 47 8p
F₁ – 10E F – 10E	☐ 14 ~ ☐ 17 4p	☐ 40 ~ ☐ 45 6p
F₂ – 12EX	☐ 14 ~ ☐ 27 12p	—
F₁ – 20E F₂ – 20E F – 20E	☐ 14 ~ ☐ 27 12p	☐ 40 ~ ☐ 47 8p
F₁ – 40E F₂ – 40E F – 40E	414 ~ 427 12p 514 ~ 527 12p	440 ~ 447 8p 540 ~ 547 8p
F₁ – 60E F₂ – 60E	014 ~ 027 12p 414 ~ 427 12p 514 ~ 527 12p	040 ~ 047 8p 440 ~ 447 8p 540 ~ 547 8p

p: points

Figure A3.2   Basic programming instructions used by Mitsubishi F series *(concluded)*.

- Analogue input/output modules available
- Plug/unplug EEPROM cassette
- Ladder diagram and instruction code programming methods are used
- Attachable graphics and key pad programming consoles
- Programmable with PC-based software (MEDOC)
- Instructions and features common to other F series PLCs.

A3.2   Programming examples

The basic programming instructions and data points used by Mitsubishi F1 series PLCs are shown in Fig. A3.2. Each instruction (e.g. LD, LDI, AND, OR, OUT, etc.) has a corresponding key on the key pad type programming console. Some examples of converting ladder diagrams into Mitsubishi instruction code are given below.

A3.2.1   LD, LDI AND OUT INSTRUCTIONS (see Fig. A3.3)

- LD is the instruction to start a logic line or branch with an N/O contact.
- LDI is the instruction to start a logic line or branch with an N/C contact.
- OUT is the instruction for an output.

A3.2.2   AND, ANI INSTRUCTIONS (see Fig. A3.4)

- AND is the instruction to connect an N/O contact in series.
- ANI is the instruction to connect an N/C contact in series.

A3.2.3   OR, ORI INSTRUCTIONS (see Fig. A3.5)

- OR is the instruction to connect an N/O contact in parallel.
- ORI is the instruction to connect an N/C contact in parallel.

A3.2.4   ORB (OR BLOCK) INSTRUCTION (See Fig. A3.6)

- ORB is the instruction for logically OR-ing two blocks.

A3.2.5   ANB (AND BLOCK) INSTRUCTION (see Fig. A3.7 on page 130)

- ANB is the instruction for logically AND-ing two blocks.

A3.2.6   TIMER (see Fig. A3.8 on page 130)

- T50 is an on-delay timer.
- The pre-set value K can take the range 0.1 to 999 s.

A3.2.7   DOWN COUNTER (see Fig. A3.9 on page 130)

- C60 is a down counter.
- The pre-set value K can take the range 1 to 999.

```
LD 400
OUT 430
LDI 401
OUT 431
```

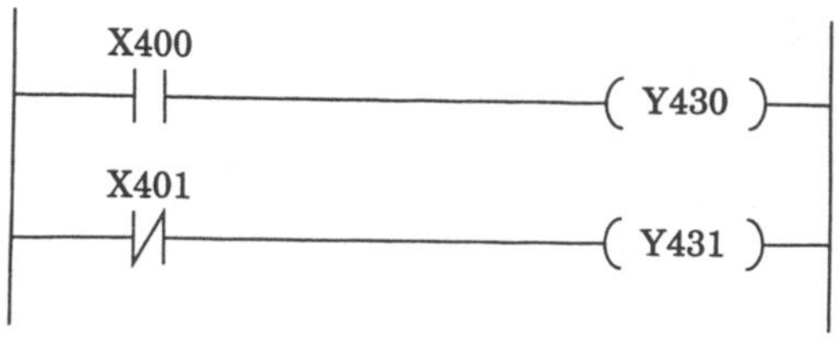

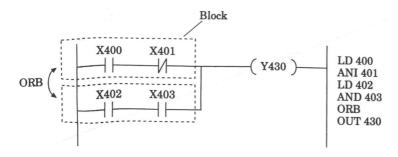

Figure A3.3 LD, LDI and OUT instructions.

```
LD 400
AND 401
OUT 430
LD 402
ANI 403
OUT 431
```

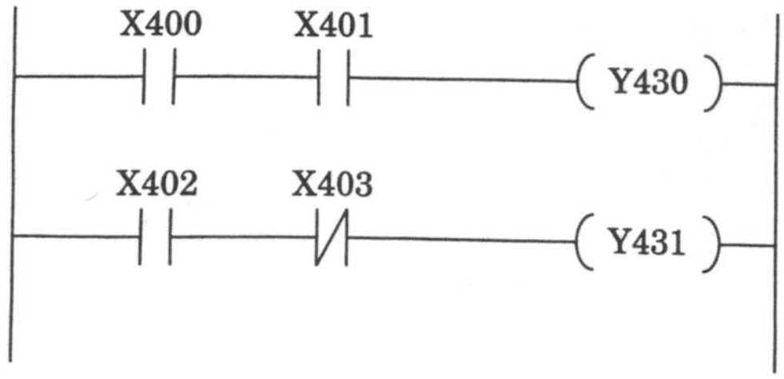

Figure A3.4 AND and ANI instructions.

```
LD 400
OR 401
OUT 430
LD 402
ORI 403
OUT 431
```

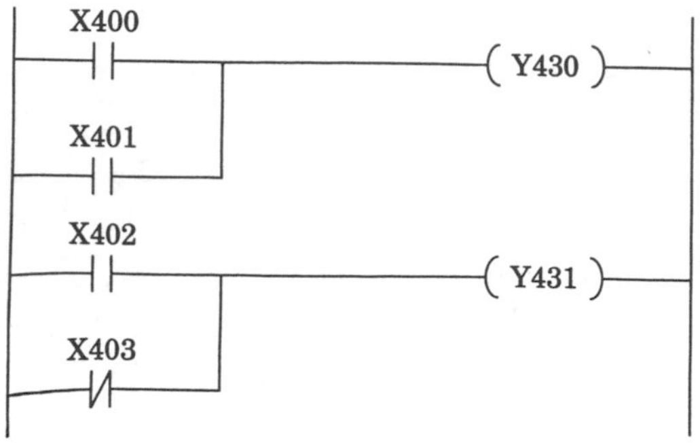

Figure A3.5 OR and ORI instructions.

```
LD 400
ANI 401
LD 402
AND 403
ORB
OUT 430
```

Figure A3.6 ORB (OR block) instruction.

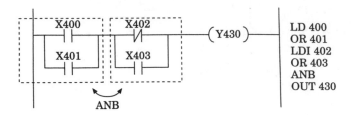

Figure A3.7   ANB (AND block) instruction.

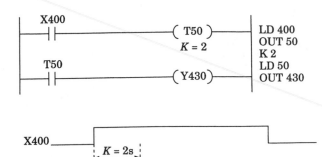

Timing diagram

Figure A3.8   On-delay timer.

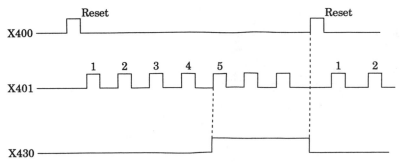

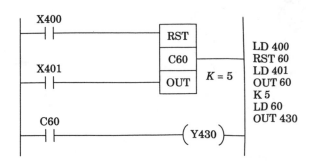

Timing diagram

Figure A3.9   Down counter.

A3.2.8   UP/DOWN COUNTER (see Fig. A3.10)

- C660 can be used as an up/down counter.
- The auxiliary relay M471 is used for up/down selection.
- The count value can be monitored using a keypad display.

A3.2.9   SHIFT REGISTER (see Fig. A3.11)

- Auxiliary relays or memory elements M100–M117 (octal) can be used as a shift register.
- Data input is to M100.
- A shift pulse shifts the data in M100–M117 one bit to the left.
- Outputs Y430–Y433 are driven by M100–M103 in Fig. A3.11.

A3.2.10   GENERATING A PULSE (see Fig. A3.12)

- A positive-going pulse is generated on memory element M102 using the pulse (PLS) function.

A3.2.11   SET AND RESET INSTRUCTIONS (see Fig. A3.13)

- S is the instruction for setting the latch.
- R is the instruction for resetting the latch.

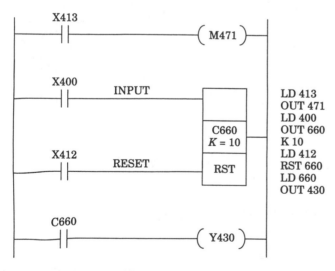

To monitor the count value in run mode using a
keypad display press the key sequence clear, 660, monitor.
M471 = ON for up counter
M471 = OFF for down counter

Figure A3.10   Up/down counter.

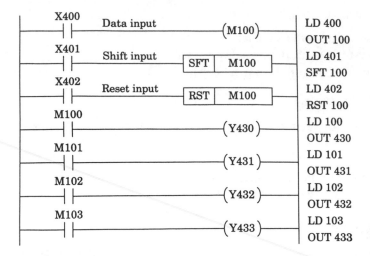

Figure A3.11   Shift register.

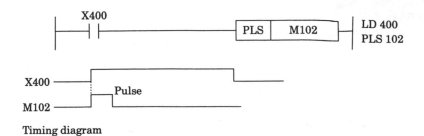

Timing diagram

Figure A3.12   Generating a pulse.

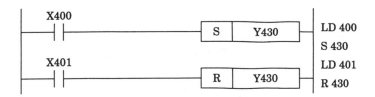

Figure A3.13   Set and reset instructions.

A3.3   Simple sequence flow

Mitsubishi supports sequential function chart (SFC) programming using a technique called 'Stepladder'. An example of a simple sequence flow is shown in Fig. A3.14.

The transitions are the states of the input contacts labelled X400, X401, X402, X403 and X404. Steps can be defined using state elements 600–647. In this

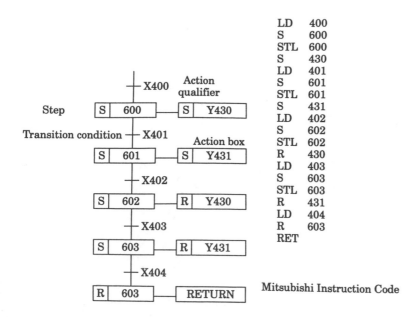

```
LD   400
S    600
STL  600
S    430
LD   401
S    601
STL  601
S    431
LD   402
S    602
STL  602
R    430
LD   403
S    603
STL  603
R    431
LD   404
R    603
RET
```

Mitsubishi Instruction Code

Figure A3.14   Sequential flow 'Stepladder'.

example states 600–603 are used. Each step has an action which either sets (e.g. latches) a coil to an ON state or resets (e.g. clears) a coil to an OFF state.

When a transition becomes TRUE (e.g. by pulsing an input contact) it causes the step before the transition to be deactivated and the step that follows the transition to be activated. Each action box has a S (SET) or R (RESET) qualifier. When the sequence is executed, the two outputs Y430 and Y431 are first latched on and then cleared. Note that the final state element 603 has to be reset and a return instruction is needed to cycle the control sequence.

A3.4   Divergent branch

The example shown in Fig. A3.15 demonstrates a divergent branch construct within a sequential function chart. Only one path is selected by testing transition conditions X401 and X402. Consequently, only one of the outputs Y431 or Y432 is selected. Note that an N (none) action qualifier has been used.

A3.5 Parallel sequences

The example shown in Fig. A3.16 demonstrates how to activate two sequences in parallel. The sequences continue independently until both the last states 603 and 605 have been reached. The step state 606 is set when both 603 and 605 are active and the transition X404 is true.

Other programming features include instructions for data transfer, data comparison and arithmetic operations.

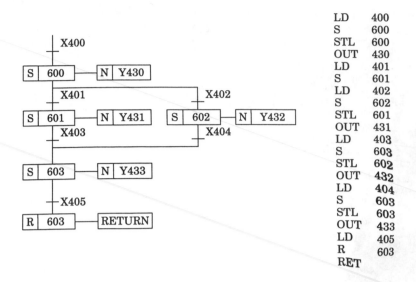

LD	400
S	600
STL	600
OUT	430
LD	401
S	601
LD	402
S	602
STL	601
OUT	431
LD	403
S	603
STL	602
OUT	432
LD	404
S	603
STL	603
OUT	433
LD	405
R	603
RET	

Mitsubishi Instruction Code

Figure A3.15   Divergent branch.

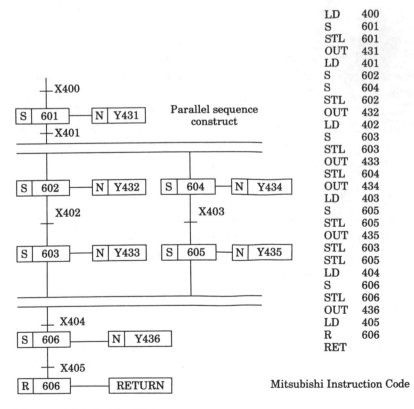

LD	400
S	601
STL	601
OUT	431
LD	401
S	602
S	604
STL	602
OUT	432
LD	402
S	603
STL	603
OUT	433
STL	604
OUT	434
LD	403
S	605
STL	605
OUT	435
STL	603
STL	605
LD	404
S	606
STL	606
OUT	436
LD	405
R	606
RET	

Mitsubishi Instruction Code

Figure A3.16   Parallel sequences.

Mitsubishi Electric UK Limited
Industrial Division
Travellers Lane
Hatfield, Hertfordshire, AL10 8XB

Appendix 4
Omron CK range PLCs

This appendix provides a brief overview of the CK range and the modular format CQM1 programmable logic controllers.

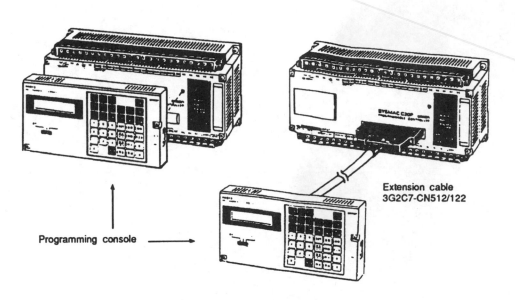

Figure A4.1 Omron CK PLC. *Source*: Sysmac Programmable Controllers C20K, C28K, C40K, C60K, Operation Manual, Omron, 1988.

A4.1 Main system features of Omron CK controllers

The Omron CK range of programmable logic controllers (see Figs. A4.1 and A4.2) is designed for use in small-to-medium scale I/O applications. The main system features are:

● Base units up to 60 I/O point capacity

- Extension units can be combined with a base unit to increase I/O capacity to 148
- LED indication of I/O status
- Simulator switches for inputs
- Analogue input/output modules available
- Programs can be transferred to EPROM
- Ladder diagram and instruction code programming methods are used
- Attachable key pad programming console
- Programmable with PC-based software (via RS232C and RS485)
- Instructions and features common to the CH and other C series PLCs

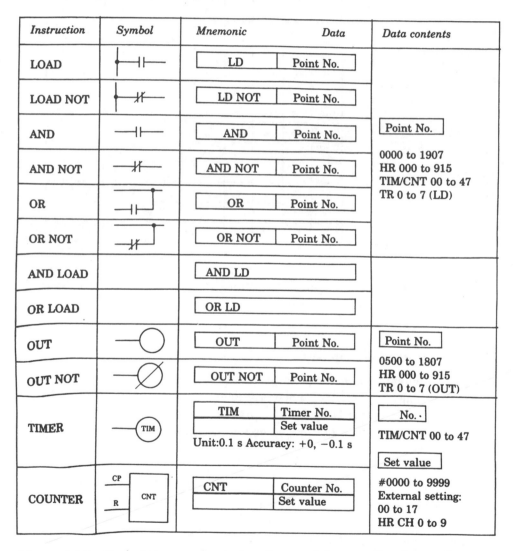

Instruction	Symbol	Mnemonic	Data	Data contents
LOAD		LD	Point No.	
LOAD NOT		LD NOT	Point No.	
AND		AND	Point No.	Point No.
AND NOT		AND NOT	Point No.	0000 to 1907 HR 000 to 915 TIM/CNT 00 to 47
OR		OR	Point No.	TR 0 to 7 (LD)
OR NOT		OR NOT	Point No.	
AND LOAD		AND LD		
OR LOAD		OR LD		
OUT		OUT	Point No.	Point No.
OUT NOT		OUT NOT	Point No.	0500 to 1807 HR 000 to 915 TR 0 to 7 (OUT)
TIMER	TIM	TIM	Timer No.	No. ·
			Set value	TIM/CNT 00 to 47
		Unit:0.1 s Accuracy: +0, −0.1 s		Set value
COUNTER	CP / R CNT	CNT	Counter No.	#0000 to 9999 External setting: 00 to 17 HR CH 0 to 9
			Set value	

Figure A4.2 Basic programming instructions and data used by Omron Sysmac PLCs C20K, C28K, C40K *(continued)*.

RELAY NUMBERS

Name	CH no. and point no.									
	CH	00	CH	01	CH	02	CH	03	CH	04
	00	08	00	08	00	08	00	08	00	08
	01	09	01	09	01	09	01	09	01	09
	02	10	02	10	02	10	02	10	02	10
	03	11	03	11	03	11	03	11	03	11
	04	12	04	12	04	12	04	12	04	12
	05	13	05	13	05	13	05	13	05	13
Holding relays	06	14	06	14	06	14	06	14	06	14
(HR0000 to 0915)	07	15	07	15	07	15	07	15	07	15
	CH	05	CH	06	CH	07	CH	08	CH	09
	00	08	00	08	00	08	00	08	00	08
	01	09	01	09	01	09	01	09	01	09
	02	10	02	10	02	10	02	10	02	10
	03	11	03	11	03	11	03	11	03	11
	04	12	04	12	04	12	04	12	04	12
	05	13	05	13	05	13	05	13	05	13
	06	14	06	14	06	14	06	14	06	14
	07	15	07	15	07	15	07	15	07	15

Figure A4.2 Basic programming instructions and data used by Omron Sysmac PLCs C20K, C28K, C40K *(continued)*.

A4.2   Programming examples

The basic programming instructions and data points used by the Omron CK range of PLCs are shown in Fig. A4.2. Each instruction (e.g. LD, NOT, AND, OR, OUT etc.) has a corresponding key on the programming console (see Fig. 2.11). Some examples of converting ladder diagrams into Omron instruction code are given below.

A4.2.1   LD, NOT AND OUT INSTRUCTIONS (see Fig. A4.3 on page 141)

● LD is the instruction to start a logic line or branch with an N/O contact.
● NOT is the instruction for inversion.
● OUT is the instruction for an output.

Name	CH no. and point no.									
	CH00 (input)		CH01 (output)		CH02 (input)		CH03 (output)		CH04 (input)	
	00	08	00	08	00	08	00	08	00	08
	01	09	01	09	01	09	01	09	01	09
	02	10	02	10	02	10	02	10	02	10
	03	11	03	11	03	11	03	11	03	11
	04	12	04	12	04	12	04	12	04	12
	05	13	05	13	05	13	05	13	05	13
	06	14	06	14	06	14	06	14	06	14
Input/output points (0000 to 0915)	07	15	07	15	07	15	07	15	07	15
	CH05 (output)		CH06 (input)		CH07 (output)		CH08 (input)		CH09 (output)	
	00	08	00	08	00	08	00	08	00	08
	01	09	01	09	01	09	01	09	01	09
	02	10	02	10	02	10	02	10	02	10
	03	11	03	11	03	11	03	11	03	11
	04	12	04	12	04	12	04	12	04	12
	05	13	05	13	05	13	05	13	05	13
	06	14	06	14	06	14	06	14	06	14
	07	15	07	15	07	15	07	15	07	15
	CH	10	CH	11	CH	12	CH	13	CH	14
	00	08	00	08	00	08	00	08	00	08
	01	09	01	09	01	09	01	09	01	09
	02	10	02	10	02	10	02	10	02	10
	03	11	03	11	03	11	03	11	03	11
	04	12	04	12	04	12	04	12	04	12
	05	13	05	13	05	13	05	13	05	13
	06	14	06	14	06	14	06	14	06	14
	07	15	07	15	07	15	07	15	07	15
	CH	15	CH	16	CH	17	CH	18		
	00	08	00	08	00	08	00			
	01	09	01	09	01	09	01			
Internal auxiliary relays (1000 to 1807)	02	10	02	10	02	10	02			
	03	11	03	11	03	11	03			
	04	12	04	12	04	12	04			
	05	13	05	13	05	13	05			
	06	14	06	14	06	14	06			
	07	15	07	15	07	15	07			

Figure A4.2 Basic programming instructions and data used by Omron Sysmac PLCs C20K, C28K, C40K *(continued)*.

Name	TIM/CNT no.					
Timer/counter (TIM/CNT00 to 47)	00	08	16	24	32	40
	01	09	17	25	33	41
	02	10	18	26	34	42
	03	11	19	27	35	43
	04	12	20	28	36	44
	05	13	21	29	37	45
	06	14	22	30	38	
	07	15	23	31	39	

Name	DM CH no.[2]							
Data memory channel (DM00 to 63)	00	08	16	24	32	40	48	56
	01	09	17	25	33	41	49	57
	02	10	18	26	34	42	50	58
	03	11	19	27	35	43	51	59
	04	12	20	28	36	44	52	60
	05	13	21	29	37	45	53	61
	06	14	22	30	38	46	54	62
	07	15	23	31	39	47		63

[1] TIM/CNT 46 and 47 are reserved for RDM and HDM, respectively.

[2] DM CHs 00–31 are reserved for RDM and DM CHs 32–62 for HDM.

Name	No. of points	Point no.	
Temporary memory relays (TR)	8	TR	0
			1
			2
			3
			4
			5
			6
			7

Note: Points 0000, 0001, and 1804 to 1807 are reserved when HDM and RDM is used.

Figure A4.2 Basic programming instructions and data used by Omron Sysmac PLCs C20K, C28K, C40K *(concluded)*.

0000
0100
0001
0101

```
LD 0000
OUT 0100
LD NOT 0001
OUT 0101
```

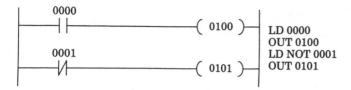

Figure A4.3 LD, NOT, and OUT instructions.

A4.2.2 AND, AND NOT INSTRUCTIONS (see Fig. A4.4)

- AND is the instruction to connect an N/O contact in series.

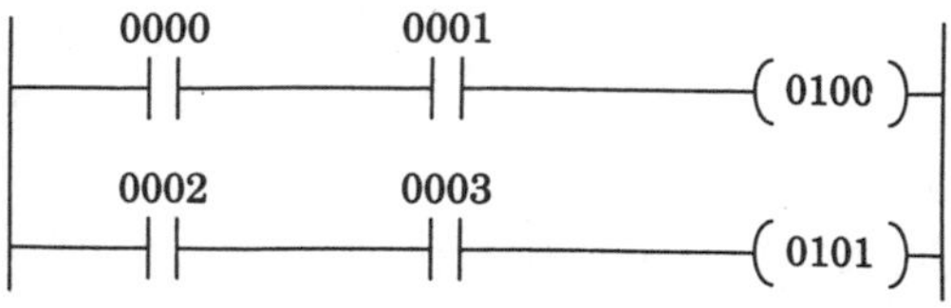

0000
0001
0100
0002
0003
0101

```
LD 0000
AND 0001
OUT 0100
LD 0002
AND NOT 0003
OUT 0101
```

Figure A4.4 AND and AND NOT instructions.

A4.2.3 OR, OR NOT INSTRUCTIONS (see Fig. A4.5)

- OR is the instruction to connect an N/C contact in parallel.

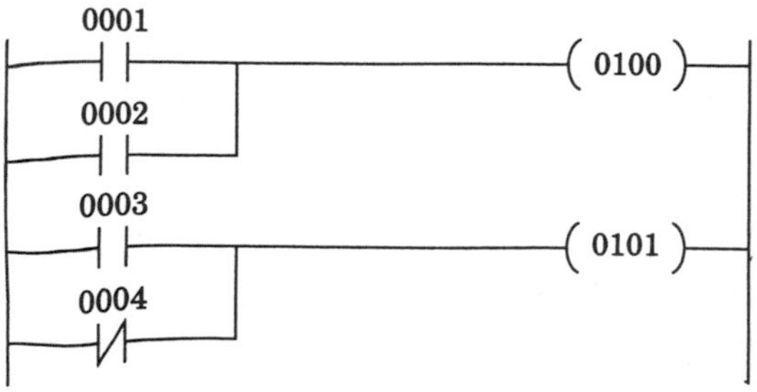

0001
0100
0002
0003
0101
0004

```
LD 0001
OR 0002
OUT 0100
LD 0003
OR NOT 0004
OUT 0101
```

Figure A4.5 OR and OR NOT instructions.

A4.2.4   OR LD INSTRUCTION (see Fig. A4.6)

● OR LD is the instruction for logically OR-ing two blocks.

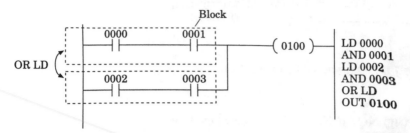

Figure A4.6   OR LD (OR block) instruction.

A4.2.5   AND LD INSTRUCTION (see Fig. A4.7)

● AND LD is the instruction for logically AND-ing two blocks.

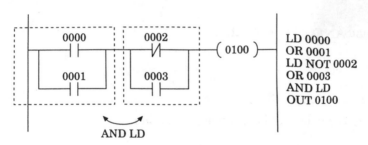

Figure A4.7   AND LD (AND block) instruction.

A4.2.6   ON-DELAY TIMER (see Fig. A4.8)

● TIM is an on-delay timer which measures in units of 0.1 s.
● The set value (SV) can take the range from 0 to 999.9 s.

```
     0000
      ┤├                  (TIM 00)      LD 0000
                           T#5s         TIM 00
     TIM 00                             # 0050
      ┤├                  ( 0100 )      LD TIM 00
                                        OUT 0100
```

Figure A4.8   On-delay timer.

A4.2.7 DOWN COUNTER (see Fig. A4.9)

- CNT is a down counter.
- The counter set value can range from 0 to 9999.

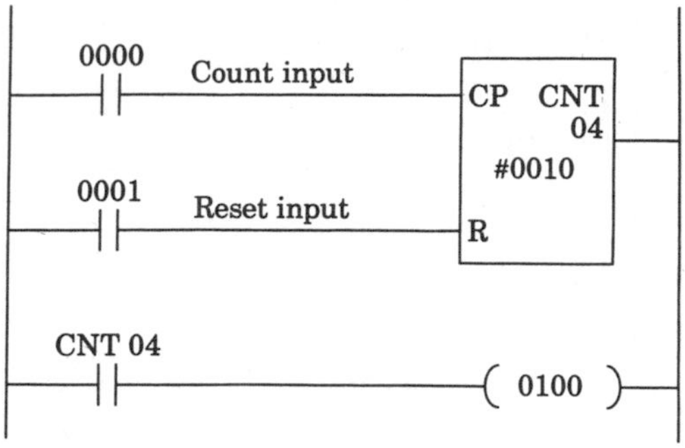

```
0000   Count input      CP   CNT
 | |                          04      LD 0000
                            #0010     LD 0001
0001   Reset input                   CNT 04
 | |                     R            # 0010
                                      LD CNT 04
                                      OUT 0100
CNT 04
 | |                      ( 0100 )
```

Figure A4.9 Down counter.

A4.2.8 REVERSIBLE COUNTER (see Fig. A4.10)

- CNTR is a reversible up–down counter.
- The counter set value can range from 0 to 9999.

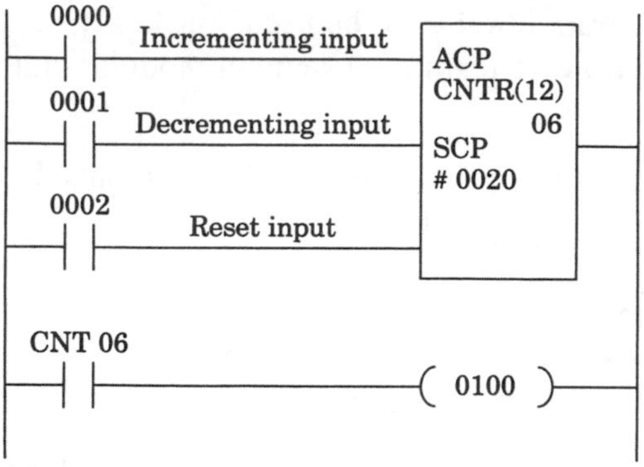

```
0000  Incrementing input   ACP
 | |                       CNTR(12)
0001  Decrementing input        06    LD 0000
 | |                       SCP         LD 0001
                           # 0020      LD 0002
0002  Reset input                      CNTR (12) 006
 | |                                   # 0020
                                       LD CNT 06
                                       OUT 0100
CNT 06
 | |                       ( 0100 )
```

Figure A4.10 Reversible up/down counter.

A4.2.9 SHIFT REGISTER (see Fig. A4.11)

- Holding relay HR9 (i.e. memory elements 900–915) acts as a 16-bit shift register.
- Data input is to HR900.
- A shift pulse shifts the data in HR9 one bit to the left.
- Output 0100 turns on when HR904 contains a logic 1.

Other programming features include high-speed and reversible drum counters, binary/BCD data manipulation, step sequencing and subroutine implementation.

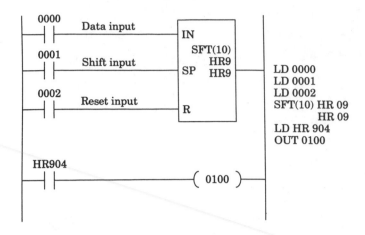

Figure A4.11   Shift register.

A4.3   Main system features of the Omron CQM1 PLC

The Omron CQM1 is a typical example of a modular PLC system in which it is possible to build a controller from a wide selection of units to match the application requirements. The CQM1 is compatible with other C-series PLCs, allowing existing programming consoles and program libraries to be used, but has a more versatile instruction set and advanced hardware features. The main modular units are:

● A range of 6 CPUs with varying program and data memory, interrupt inputs, I/O points and built-in features for a variety of applications
● Three power supply units
● Input units with d.c. inputs and a.c. inputs
● Output units with relay contact, transistor and triac output stages
● Dedicated I/O units for specific application tasks such as temperature control.

The basic specification of the CQM1 modular units currently available are illustrated in Fig. A4.12.

The CPU modules CPU11-E and CPU21-E are referred to as 'standard CPUs' while the others are referred to as 'high-performance high-capacity CPUs'. All CPU modules except the CPU11-E incorporate a built-in RS232C port for connecting external serial devices such as computers, printers and programmable terminals. The CPU43-E module has two high-speed pulse I/O ports, while the CPU44-E has a port for reading an absolute encoder. All CPU modules support interrupt input processing, important for high-speed I/O response applications.

Input Units

Input Units with DC inputs or AC inputs and capacities from 8 to 32 points are available.

CQM1-ID211   CQM1-ID212   CQM1-ID213   CQM1-IA121

Inputs	Input points	Input voltage	Configuration	Model
DC	8	12 to 24 VDC	Independent contacts	CQM1-ID211
	16	24 VDC	16 points/common	CQM1-ID212
	32		32 points/common	CQM1-ID213
AC	8	100 to 120 VAC	8 points/common	CQM1-IA121
		200 to 240 VAC		CQM1-IA221

Output Units

Output Units with contact, transistor, and triac outputs, and capacities from 8 to 32 points are available.

CQM1-OC221   CQM1-OD211   CQM1-OD213   CQM1-OA221

Outputs	Output points	Max. switching voltage	Configuration	Model
Contact	8	250 VAC/ 24 VDC	Independent contacts	CQM1-OC221
	16		16 pts/com	CQM1-OC222
Transistor	8	24 VDC	8 pts/com	CQM1-OD211
	16		16 pts/com	CQM1-OD212
	32		32 pts/com	CQM1-OD213
	16	24 VDC	16 pts/com	CQM1-OD214
	8	PNP	8 pts/com	CQM1-OD215
AC	8	100 to 240 VAC	4 pts/com 2 circuits	CQM1-OA221

Dedicated I/O Units

● **CQM1-AD041**
Analog Input Unit
This Unit can input 4 analog voltage or current signals into the CQM1.

● **CQM1-DA021**
Analog Output Unit
This Unit allows two-point digital-to-analog conversion.

● **CQM1-IPS01/02**
Power Supply Unit
This Unit is used to supply power when using an Analog Input or Output Unit.

● **CQM1-B7A□□**
B7A Interface Unit
This Unit allows an equal number of points to be connected to 16-point B7A Link Terminals.

● **CQM1-TC00□/10□**
Temperature Control Unit
A single Temperature Control Unit connects to two temperature control loops and are ideal for simple ON and OFF control.

● **CQM1-LSE0□**
Linear Sensor Interface Unit
Convert voltage or current inputs from linear sensors quickly and accurately to numeric data for comparative decision processing.

● **CQM1-LK501**
I/O Link Unit
This Unit allows the CQM1 to be a part of the SYSMAC BUS Wired Remote I/O System.

● **CQM1-SEN01**
Sensor Unit
The CQM1-SEN01 Sensor Unit is a space-saving model that reduces wiring effort and allows the direct connection of various sensors to the CQM1.

● **CQM1-G7M21/G7N□1**
G730 Interface Unit
A G730 Interface Unit allows signals for remote I/O equipment to be handled and controlled by the G730 Remote Terminal.

Figure A4.12   CQM1 modules. *Source*: Omron CQM1 operations manual, (cat no. P31-E1-5), Omron, 1996 *(concluded)*.

A4.3.1 CQM1 PROGRAMMING INSTRUCTIONS

There are 117 instructions for the standard CPU units (CPU11-E and CPU21-E) and 137 instructions for the high-performance CPUs. A summary of the programming instructions is shown in Fig. A4.13 and describes the software functionality of the system. Noteworthy are the advanced I/O instructions for reading key pads and outputing BCD data to a seven-segment display.

● **Basic Instruction Set**

Code	Instruction	Mnem.	Function
---	LOAD	LD	Connects an NO condition to the left bus bar.
---	LOAD NOT	LD NOT	Connects an NC condition to the left bus bar.
---	AND	AND	Connects an NO condition in series with the previous condition.
---	AND NOT	AND NOT	Connects an NC condition in series with the previous condition.
---	OR	OR	Connects an NO condition in parallel with the previous condition.
---	OR NOT	OR NOT	Connects an NC condition in parallel with the previous condition.
---	AND LOAD	AND LD	Connects two instruction blocks in series.
---	OR LOAD	OR LD	Connects two instruction blocks in parallel.
---	OUTPUT	OUT	Outputs the result of logic to a bit.
---	OUT NOT	OUT NOT	Reverses and outputs the result of logic to a bit.
---	SET	SET	Force sets (ON) a bit.
---	RESET	RSET	Force resets (OFF) a bit.
---	COUNTER	CNT	A decrementing counter.
12	REVERSIBLE COUNTER	CNTR	Increases or decreases PV by one.
---	TIMER	TIM	An ON-delay (decrementing) timer.
15	HIGH-SPEED TIMER	TIMH	A high-speed, ON-delay (decrementing) timer.
01	END	END	Required at the end of the program.
02	INTERLOCK	IL	If the execution condition for IL(02) is OFF, all outputs are turned OFF and all timer PVs reset between IL(02) and the next ILC(03).
03	INTERLOCK CLEAR	ILC	ILC(03) indicates the end of an interlock (beginning at IL(02)).
04	JUMP	JMP	If the execution condition for JMP(04) is ON, all instructions between JMP(04) and JME(05) are treated as NOP(00).
05	JUMP END	JME	JME(05) indicates the end of a jump (beginning at JMP(04))
11	KEEP	KEEP	Maintains the status of the designated bit.
13	DIFFERENTIATE UP	DIFU	Turns ON a bit for one scan when the execution condition goes from OFF to ON.
14	DIFFERENTIATE DOWN	DIFD	Turns ON a bit for one scan when the execution condition goes from ON to OFF.

● **Advanced I/O Instructions**

Code	Instruction	Mnem.	Function
18	TEN KEY INPUT	TKY	Inputs 8 digits of BCD data from a 10-key keypad.
---*	HEXADECIMAL KEY INPUT	HKY	Up to 8 lines of hexadecimal data can be retrieved using a 16-key keypad.
87	DIGITAL SWITCH	DSW	Inputs 4 or 8-digit BCD data from a digital switch.
88	7-SEGMENT DISPLAY OUTPUT	7SEG	Converts 4 or 8-digit BCD data to 7-segment display format and then outputs the converted data.

● **Data Comparison Instructions**

Code	Instruction	Mnem.	Function
20	COMPARE	CMP	Compares two four-digit hexadecimal values.
---*	SIGNED BINARY COMPARE	CPS	Compares word data values in signed four-digit binary.
60	DOUBLE COMPARE	CMPL	Compares two eight-digit hexadecimal values.
---*	SIGNED BINARY DOUBLE COMPARE	CPSL	Compares two word values in signed eight-digit binary.
68	BLOCK COMPARE	(@)BCMP	Judges whether the value of a word is within 16 ranges (defined by lower and upper limits).
85	TABLE COMPARE	(@)TCMP	Compares the value of a word to 16 consecutive words.
19	MULTI-WORD COMPARE	(@)MCMP	Compares a block of 16 consecutive words to a block of 16 consecutive words.
---*	RANGE COMPARE	ZCP	Checks if the specified word data exists between the upper limit and lower limit values specified by the four-digit binary value.
---*	RANGE DOUBLE COMPARE	ZCPL	Checks if the specified two-word data exists between the upper limit and lower limit values specified by the eight-digit binary value.

● **Differentiated Instructions**
- An instruction marked with "@" can be used as a differentiated instruction that will be executed only once each time the instruction executing condition is turned ON.
- Dedicated Instruction for CPU41-E/42-E/43-E/44-E

 Instructions within colored boxes (☐) are supported by the CPU41-E, CPU42-E, CPU43-E, and CPU44-E only.

- Asterisk-marked function codes are for expansion instructions. Before using these function instructions, exchange standard instructions with them.

Figure A4.13   Programming instructions. *Source*: Omron CQM1 operations manual, (cat no. P31-E1-5), Omron, 1996 *(continued)*.

● **Data Movement Instructions**

Code	Instruction	Mnem.	Function
21	MOVE	(@)MOV	Copies a constant or the content of a word to a word.
22	MOVE NOT	(@)MVN	Copies the complement of a constant or the content of a word to a word.
70	BLOCK TRANS-FER	(@)XFER	Copies the content a block of up to 2000 consecutive words to a block of consecutive words.
73	DATA EXCHANGE	(@)XCHG	Exchanges the content of two words.
71	BLOCK SET	(@)BSET	Copies the content of a word to a block of consecutive words.
82	MOVE BIT	(@)MOVB	Copies the specified bit from one word to the specified bit of a word.
—*	MULTIPLE BIT TRANSFER	(@)XFRB	Transfers consecutive bit data values.
83	MOVE DIGIT	(@)MOVD	Copies the specified digits (4-bit units) from a word to the specified digits of a word.
80	SINGLE WORD DISTRIBUTE	(@)DIST	Copies the content of a word to a word (whose address is determined by adding an offset to a word address).
81	DATA COLLECT	(@)COLL	Copies the content of a word (whose address is determined by adding an offset to a word address) to a word.

● **Shift Instructions**

Code	Instruction	Mnem.	Function
10	SHIFT REGISTER	SFT	Copies the specified bit (0 or 1) into the rightmost bit of a shift register and shifts the other bits one bit to the left.
84	REVERSIBLE SHIFT REGISTER	(@)SFTR	Creates a single or multiple-word shift register that can shift data to the left or right.
17	ASYNCHRO-NOUS SHIFT REGISTER	(@)ASFT	Creates a shift register that exchanges the contents of adjacent words when one is zero and the other is not.
16	WORD SHIFT	(@)WSFT	Creates a multiple-word shift register that shifts data to the left in one-word units.
25	ARITHMETIC SHIFT LEFT	(@)ASL	Shifts a 0 into bit 00 of the specified word and shifts the other bits one bit to the left.
26	ARITHMETIC SHIFT RIGHT	(@)ASR	Shifts a 0 into bit 15 of the specified word and shifts the other bits one bit to the right.
27	ROTATE LEFT	(@)ROL	Moves the content of CY into bit 00 of the specified word, shifts the other bits one bit to the left, and moves bit 15 to CY.
28	ROTATE RIGHT	(@)ROR	Moves the content of CY into bit 15 of the specified word, shifts the other bits one bit to the right, and moves bit 00 to CY.
74	ONE DIGIT SHIFT LEFT	(@)SLD	Shifts a 0 into the rightmost digit (4-bit unit) of the shift register and shifts the other digits one digit to the left.
75	ONE DIGIT SHIFT RIGHT	(@)SRD	Shifts a 0 into the leftmost digit (4-bit unit) of the shift register and shifts the other digits one digit to the right.

● **Data Conversion Instructions**

Code	Instruction	Mnem.	Function
23	BCD TO BINARY	(@)BIN	Converts 4-digit BCD data to 4-digit binary data.
58	DOUBLE BCD TO DOUBLE BINARY	(@)BINL	Converts 8-digit BCD data to 8-digit binary data.
24	BINARY TO BCD	(@)BCD	Converts 4-digit binary data to 4-digit BCD data.
59	DOUBLE BINARY TO DOUBLE BCD	(@)BCDL	Converts 8-digit binary data to 8-digit BCD data.
67	BIT COUNTER	(@)BCNT	Counts the total number of bits that are ON in the specified block of words.
76	4 TO 16 DECODER	(@)MLPX	Takes the hexadecimal value of the specified digit(s) in a word and turns ON the corresponding bit in a word(s).
77	16 TO 4 ENCODER	(@)DMPX	Identifies the highest ON bit in the specified word(s) and moves the hexadecimal value(s) corresponding to its location to the specified digit(s) in a word.
78	7-SEGMENT DECODER	(@)SDEC	Converts the designated digit(s) of a word into an 8-bit, 7-segment display code.
86	ASCII CODE CONVERT	(@)ASC	Converts the designated digit(s) of a word into the equivalent 8-bit ASCII code.
—*	2'S COMPLE-MENT CON-VERT	(@)NEG	Takes the 2's complement of the 4-digit binary data of the designated word.
—*	2'S COMPLE-MENT DOUBLE CONVERT	(@)NEGL	Takes the 2's complement of the 8-digit binary data of the designated word.
—*	ASCII TO HEX-ADECIMAL	(@)HEX	Converts 16-bit ASCII data to hexadecimal data.
—*	COLUMN TO LINE	(@)LINE	Reads the content (1 or 0) of each bit of consecutive 16 words and converts them to word data.
—*	LINE TO COL-UMN	(@)COLM	Outputs the content (1 or 0) of each bit of the specified word to the corresponding bits of consecutive 16 words.
—*	HOURS TO SE-CONDS	(@)SEC	Converts hour and minute data to second data.
—*	SECONDS TO HOURS	(@)HMS	Converts second data to hour and minute data.
—*	ARITHMETIC PROCESS	(@)APR	Performs sine, cosine, or polygonal line approximation calculations.

Figure A4.13 Programming instructions. *Source*: Omron CQM1 operations manual, (cat no. P31-E1-5), Omron, 1996 *(continued)*.

● BCD Calculation Instructions

Code	Instruction	Mnem.	Function
30	BCD ADD	(@)ADD	Adds the content of a word (or constant).
54	DOUBLE BCD ADD	(@)ADDL	Adds the 8-digit BCD contents of two pairs of words (or constants) and CY.
31	BCD SUBTRACT	(@)SUB	Subtracts the content of a word (or constant) and CY from the content of a word (or constant).
55	DOUBLE BCD SUBTRACT	(@)SUBL	Subtracts the 8-digit BCD content of a pair of words (or constant) and CY from the 8-digit BCD content of a pair of words (or constant).
32	BCD MULTIPLY	(@)MUL	Multiplies the contents of two words (or constants).
56	DOUBLE BCD MULTIPLY	(@)MULL	Multiplies the 8-digit BCD contents of two pairs of words (or constants).
33	BCD DIVIDE	(@)DIV	Divides the content of a word (or constant) with the content of a word (or constant).
57	DOUBLE BCD DIVIDE	(@)DIVL	Divides the 8-digit BCD content of a pair of words (or constant) with the 8-digit BCD content of a pair of words (or constant).
40	SET CARRY	(@)STC	Set Carry Flag 25504 to 1.
41	CLEAR CARRY	(@)CLC	Set Carry Flag 25504 to 0.
38	INCREMENT	(@)INC	Increments the BCD content of the specified word by 1.
39	DECREMENT	(@)DEC	Decrements the BCD content of the specified word by 1.
72	SQUARE ROOT	(@)ROOT	Computes the square root of the 8-digit BCD content of a pair of words (or constant).

● Binary Calculation Instructions

Code	Instruction	Mnem.	Function
50	BINARY ADD	(@)ADB	Adds the contents of two words (or constants) and CY.
—*	BINARY DOUBLE ADD	(@)ADBL	Adds the contents of two words or constants (eight digits) in binary.
51	BINARY SUBTRACT	(@)SBB	Subtracts the content of a word (or constant) and CY from the content of a word (or constant).
—*	BINARY DOUBLE SUBTRACT	(@)SBBL	Subtracts the content of a word or constant (eight digits) from the content of a word or constant (eight digits) in binary.
52	BINARY MULTIPLY	(@)MLB	Multiplies the contents of two words (or constants).
—*	SIGNED BINARY MULTIPLY	(@)MBS	Multiplies the contents of two words or constants in signed binary.
—*	SIGNED BINARY DOUBLE MULTIPLY	(@)MBSL	Multiplies the contents of two words or constants (eight digits) in signed binary.

Code	Instruction	Mnem.	Function
53	BINARY DIVIDE	(@)DVB	Divides the content of a word (or constant) by the content of a word and obtains the result and remainder.
—*	SIGNED BINARY DIVIDE	(@)DBS	Divides the content of a word or constant with the content of a word or constant in signed binary.
—*	SIGNED BINARY DOUBLE DIVIDE	(@)DBSL	Divides the content of a word or constant (eight digits) with the content of a word or constant in signed binary.

● Logic Instructions

Code	Instruction	Mnem.	Function
34	LOGICAL AND	(@)ANDW	Logically ANDs the corresponding bits of two words (or constants).
35	LOGICAL OR	(@)ORW	Logically ORs the corresponding bits of two words (or constants).
36	EXCLUSIVE OR	(@)XORW	Exclusively ORs the corresponding bits of two words (or constants).
37	EXCLUSIVE NOR	(@)XNRW	Exclusively NORs the corresponding bits of two words (or constants).
29	COMPLEMENT	(@)COM	Turns OFF all ON bits and turns ON all OFF bits in the specified word.

● Subroutine Instructions

Code	Instruction	Mnem.	Function
91	SUBROUTINE ENTER	(@)SBS	Executes a subroutine in the main program.
92	SUBROUTINE ENTRY	SBN	Marks the beginning of a subroutine program.
93	SUBROUTINE RETURN	RET	Marks the end of a subroutine program.
99	MACRO	MCRO	Calls and executes the specified subroutine, substituting the specified input and output words for the input and output words in the subroutine.
89	INTERRUPT CONTROL	(@)INT	Performs interrupt control, such as masking and unmasking the interrupt bits for I/O interrupts.
69	INTERVAL TIMER	(@)STIM	Controls interval timers used to perform scheduled interrupts.

● Step Instructions

Code	Instruction	Mnem.	Function
08	STEP DEFINE	STEP	Defines the start of a new step and resets the previous step when used with a control bit. Defines the end of step execution when used without a control bit.
09	STEP START	SNXT	Starts the execution of the step when used with a control bit.

Figure A4.13 Programming instructions. *Source*: Omron CQM1 operations manual, (cat no. P31-E1-5), Omron, 1996 *(continued)*.

● Special Processing Instructions

Code	Instruction	Mnem.	Function
45	TRACE MEMORY SAMPLE	TRSM	Marks locations in the program where specified data will be sampled and stored in Trace Memory.
46	MESSAGE	(@)MSG	Reads up to 8 words of ASCII code (16 characters) from memory and displays the message on the Programming Console or other Peripheral Device.
61	MODE CONTROL	(@)INI	Starts and stops counter operation, compares and changes counter PVs, and stops pulse output.
62	PV READ	(@)PRV	Reads counter PVs and status data.
63	COMPARE TABLE LOAD	(@)CTBL	Compares counter PVs and generates a direct table or starts operation.
64	CHANGE FREQUENCY	(@)SPED	Outputs pulses at the specified frequency (10 Hz to 50 KHz in 10 Hz units). The output frequency can be changed while pulses are being output.
---*	FREQUENCY CONTROL	(@)ACC	Controls the pulse output frequency of CPU's pulse output port 1 and 2 (CQM1-CPU43-E only)
65	SET PULSE	(@)PULS	Outputs the specified number of pulses at the specified frequency. The pulse output cannot be stopped until the specified number of pulses have been output.
66	SCALE	(@)SCL	Performs a scaling conversion on the calculated value.
---*	SCALE 2	(@)SCL2	Performs a scaling conversion on the hexadecimal data of the specified word into BCD data.
---*	SCALE 3	(@)SCL3	Performs a scaling conversion on the BCD data of the specified word into hexadecimal data.
---*	DATA SEARCH	(@)SRCH	Searches the table for any data identical to the input data.
---*	FIND MAXIMUM	(@)MAX	Finds the maximum value in the specified data area and outputs that value to the specified word.
---*	FIND MINIMUM	(@)MIN	Finds the minimum value in the specified data area and outputs that value to the specified word.
---*	SUM CALCULATE	(@)SUM	Computes the sum of the specified data table.
---*	FCS CALCULATE	(@)FCS	Checks for errors in data transmitted by a SYSMAC WAY command.
---*	AVERAGE VALUE	AVG	Adds the specified number of hexadecimal words and computes the mean value. Rounds off to 4 digits past the decimal point.
---*	FAILURE POINT DETECT	FPD	Finds errors within an instruction block.
---*	PWM OUTPUT	(@)PWM	Changes the width rate of pulse output within a range of 0% to 99% (CQM1-CPU43-E only)

Code	Instruction	Mnem.	Function
---*	POSITIONING	(@)PLS2	Performs the frequency control of the acceleration and deceleration of the specified pulse value with the same cycle in simple positioning mode (CQM1-CPU43-E only).
---*	PID CONTROL	PID	Performs PID control based on the operand and PID parameters that have been set.
---*	COMMUNICATION PORT OUTPUT	(@)TXD	Reads the specified number of bytes of the data received by the specified port.
---*	COMMUNICATION PORT INPUT	(@)RXD	Transmits the specified number of data bytes from the specified port.

● Special System Instructions

Code	Instruction	Mnem.	Function
06	FAILURE ALARM	(@)FAL	Generates a non-fatal error when executed. The Error/Alarm indicator flashes and the CPU continues operating.
07	SEVERE FAILURE ALARM	FALS	Generates a fatal error when executed. The Error/Alarm indicator lights and the CPU stops operating.
97	I/O REFRESH	(@)IORF	Refreshes the specified I/O words.

---* Before using any expansion instruction (marked with an asterisk), it is necessary to set the function number for the expansion instruction. Expansion instructions can be replaced with 18 standard instructions by using a support tool such as the SSS or Programming Console.

Figure A4.13   Programming instructions. *Source*: Omron CQM1 operations manual, (cat no. P31-E1-5), Omron, 1996 *(concluded)*.

A4.3.2   CQM1 CPU APPLICATIONS

Some typical application examples of using CPU modules are illustrated in Fig. A4.14. The first example (Fig. A4.14(a)) shows how the pulse I/O port is used for two-axis motor speed control and positioning. Figure A4.14(b) shows how a rotary table can be controlled using an absolute shaft encoder.

a) **Large-capacity CPU Incorporating RS-232C Port and Pulse I/O Function————CPU43-E**

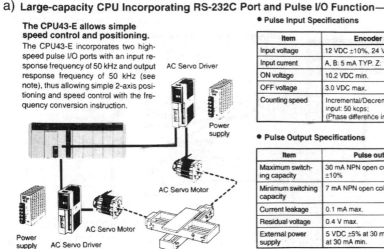

The CPU43-E allows simple speed control and positioning.
The CPU43-E incorporates two high-speed pulse I/O ports with an input response frequency of 50 kHz and output response frequency of 50 kHz (see note), thus allowing simple 2-axis positioning and speed control with the frequency conversion instruction.

● Pulse Input Specifications

Item	Encoder inputs A, B and Z	
Input voltage	12 VDC ±10%, 24 VDC ±10%	
Input current	A, B: 5 mA TYP. Z: 12 mA TYP.	
ON voltage	10.2 VDC min.	20.4 VDC min.
OFF voltage	3.0 VDC max.	4.0 VDC max.
Counting speed	Incremental/Decremental arithmetic operation input: 50 kcps; (Phase difference input: 25 kcps)	

● Pulse Output Specifications

Item	Pulse output CW and CCW
Maximum switching capacity	30 mA NPN open collector at 5 to 24 VDC ±10%
Minimum switching capacity	7 mA NPN open collector at 5 to 24 VDC ±10%
Current leakage	0.1 mA max.
Residual voltage	0.4 V max.
External power supply	5 VDC ±5% at 30 mA min.; 24 VDC +10%/−15% at 30 mA min.

Note: The maximum output frequency is 50 kHz except for stepping motor trapezoidal speed control, the maximum output frequency of which is 20 kHz.

Figure A4.14   CPU applications (a) large-capacity CPU incorporating RS-232C port and pulse I/O function—CPU43-E. *Source*: Omron CQM1 operations manual (cat no. P31-E1-5), Omron, 1996 *(continued)*.

b) Large-capacity CPU Incorporating RS-232C Port and Absolute Interface————————CPU44-E

The CPU44-E can retrieve position data from an ABS-type (absolute-type) encoder.
ABS input data is a 12-bit gray code. The CPU44-E holds positioning data at the time of power failure. This means origin reset is not necessary when the power is turned on. The CPU44-E incorporates an origin reset function, with which any position can be treated as the origin.

• **Operation Mode**

There is a BCD mode and 360° mode, either of which can be selected.

• **Resolution**

Either a resolution of 8 bits (0 to 255), 10 bits (0 to 1023), or 12 bits (0 to 4095) can be selected according to the resolution of the encoder to be connected to the CQM1.

• **Input Specifications**

Input voltage	24 VDC ±10%/–15%
Input impedance	5.4 kΩ
Input current	4 mA (TYP.)
ON voltage	16.8 VDC min.
OFF voltage	3.0 VDC max.
Counting speed	4 kHz max.
Input code	Gray code binary (8, 10, or 12 bits)

Recommended encoder: OMRON's E6CP-AG5C (8 bits at 12 to 24 VDC).

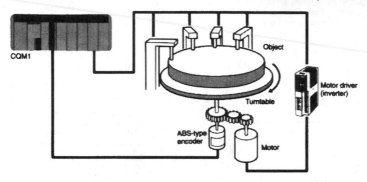

Figure A4.14   CPU applications (b) large-capacity CPU incorporating RS-232C port and pulse I/O function—CPU44-E. *Source*: Omron CQM1 operations manual (cat no. P31-E1-5), Omron, 1996 *(concluded)*.

Omron Electronics
1 Apsley Way
Staples Corner
London, NW2 7HF

Appendix 5
Toshiba Prosec T1 PLC

♦ **T1-40**

Power supply and
input terminals — Input status LEDs

Mounting
hole

IN

Option card
slot

PROSEC

Programmer
port cover

TOSHIBA **T1** MDR40

OUT

Expansion
connector

Output terminals — Operation status LEDs

Output status LEDs

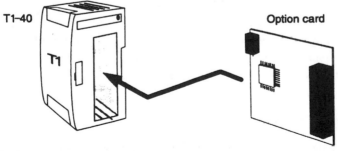
T1-40   Option card

T1

Figure A5.1 Toshiba T1-40 PLC.

A5.1 Main system features

The Prosec T1 is a high-performance block style PLC which is a member of Toshiba's T-series family of controllers (see Fig. A5.1). The main system features are:

- Base units with a range of 16–40 I/O capacity:
    T1–16: 8 input and 8 output points
    T1–28: 14 input and 14 output points
    T1–40: 24 input and 16 output points
- Control up to 200 points using option cards and expansion modules fitted to T1–40
- Built-in high speed counter
- Pulse output for driving stepper motors or
- PWM output which can be used to simulate an analogue output
- High-speed processing using an immediate interrupt function which allows:
    Inputs to be recognized independent of program scan
    Outputs turned on/off independent of program scan
- Ladder diagram and function block programming methods are used
- User programs can be stored in a built-in EEPROM
- Built-in RS232 port
- T1–40 model has two option card slots for I/O expansion and networking
- Instructions and features are common with other T-series PLCs (e.g. T2 and T3)

A5.2 Programming examples

The basic programming instructions and data points used by the Toshiba T1 range of PLCs are shown in Fig. A5.2. User programs in the form of ladder diagrams are developed using either

- T-series program development system (T-PDS) which runs on an IBM-PC compatible personal computer or
- Hand-held graphic programmer (see Fig. 2.11)

A5.2.1 FUNCTION INSTRUCTIONS

In addition to the basic instruction set which is used for developing general purpose ladder diagrams, the Prosec T1 has a large and varied function instruction set. The function instructions are incorporated within the ladder diagram itself. An example ladder program which incorporates the 'greater than' function instruction is shown in Fig. A5.3 on page 157.

A5.2.2 FUNCTION BLOCKS

Function blocks can be incorporated within the ladder diagram. For example, an up-down counter is implemented by drawing the network shown in Fig. A5.4 on page 158.

Basic instructions

FUN No.	Symbol	Name
	⊣ ⊢	NO contact
	⊣/⊢	NC contact
	⊣↑⊢	Transitional contact (rising edge)
	⊣↓⊢	Transitional contact (falling edge)
	─( )─	Coil
	×( )─	Forced coil (debugging purpose only)
	⊣ I ⊢	Inverter
	─(I)─	Invert coil
	MCS	Master control set
	MCR	Master control reset
	JCS	Jump control set
	JCR	Jump control reset
	TON	ON delay timer
	TOF	OFF delay timer
	SS	Single-shot timer
	CNT	Counter
	END	End

Function instructions

FUN No.	Symbol	Name
18	MOV	Data transfer
20	NOT	Invert transfer
22	XCHG	Exchange
27	+	Addition
28	–	Subtraction
29	*	Multiplication
30	/	Division
35	+C	Addition with carry
36	–C	Subtraction with carry
43	+1	Increment
45	–1	Decrement
48	AND	AND
50	OR	OR
52	EOR	Exclusive OR
64	TEST	Bit test
68	SHR1	1 bit shift right
69	SHL1	1 bit shift left
70	SHRn	n bits shift right
71	SHLn	n bits shift left
74	SR	Shift register
75	DSR	Bidirectional shift register

FUN No.	Symbol	Name
78	RTR1	1 bit rotate right
79	RTL1	1 bit rotate left
80	RTRn	n bits rotate right
81	RTLn	n bits rotate left
90	MPX	Multiplexer
91	DPX	Demultiplexer
96	>	Greater than
97	>=	Greater than or equal
98	=	Equal
99	<>	Not equal
100	<	Less than
101	<=	Less than or equal
114	SET	Device/register set
115	RST	Device/register reset
118	SETC	Set carry
119	RSTC	Reset carry
120	ENC	Encode
121	DEC	Decode
128	CALL	Subroutine call
129	RET	Subroutine return
132	FOR	FOR–NEXT loop (FOR)
133	NEXT	FOR–NEXT loop (NEXT)
137	SUBR	Subroutine entry
140	EI	Enable interrupt
141	DI	Disable interrupt
142	IRET	Interrupt return
143	WDT	Watchdog timer reset
144	STIZ	Step sequence initialization
145	STIN	Step sequence input
146	STOT	Step sequence output
147	F/F	Flip flop
149	U/D	Up/down counter
165	FG	Function generator
180	ABS	Absolute value
182	NEG	Two's complement
185	7SEG	Seven-segment decode
186	ASC	ASCII conversion
188	BIN	Binary conversion
190	BCD	BCD conversion
235	I/O	Direct input/output
236	XFER	Expanded data transfer
237	READ	Special module data read
238	WRITE	Special module data write
	PID 3	Proportional integral derivative (available soon)

Figure A5.2   Instruction set and data used by Prosec T1 series PLCs *(continued)*.

Available address range

Device / register	Symbol	Number of points	Address range
External input device	X	Total 512 points	X000-X31F
External output device	Y		Y020-Y31F
External input register	XW	Total 32 words	XW00-XW31
External output register	YW		YW02-YW31
Auxiliary relay device	R	1024 points	R000-R63F
Auxiliary relay register	RW	64 words	RW00-RW63
Special device	S	1024 points	S000-S63F
Special register	SW	64 words	SW00-SW63
Timer device	T.	64 points	T.00-T.63
Timer register	T	64 words	T00-T63
Counter device	C.	64 points	C.00-C.63
Counter register	C	64 words	C00-C63
Data register	D	1024 words	D0000-D1023
Index register	I	1 word	I (no address)
	J	1 word	J (no address)
	K	1 word	K (no address)

Addressing devices

A device number of X, Y, R and S devices consists of a register number and bit position as follows:

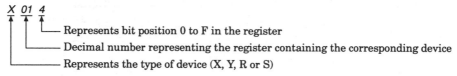

- Represents bit position 0 to F in the register
- Decimal number representing the register containing the corresponding device
- Represents the type of device (X, Y, R or S)

As for the timer (T.) and the counter (C.) devices, a device number is expressed as follows:

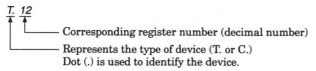

- Corresponding register number (decimal number)
- Represents the type of device (T. or C.)
  Dot (.) is used to identify the device.

Addressing registers

A register number except the index registers is expressed as follows:

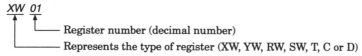

- Register number (decimal number)
- Represents the type of register (XW, YW, RW, SW, T, C or D)

The index registers (I, J and K) do not have the number.

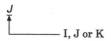

- I, J or K

Figure A5.2   Instruction set and data used by Prosec T1 series PLCs *(continued)*.

I/O allocation

The external input signals are allocated to the external input devices/registers (X/XW).
The external output signals are allocated to the external output devices/registers (Y/YW).
The register numbers of the external input and output registers are consecutive.
Thus one register number can be assigned for either input or output.

As for the T1 basic unit, I/O allocation is fixed as follows:

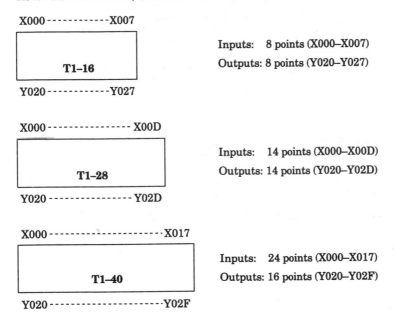

X000 ----------- X007

T1–16

Y020 ----------- Y027

Inputs: 8 points (X000–X007)
Outputs: 8 points (Y020–Y027)

X000 --------------- X00D

T1–28

Y020 --------------- Y02D

Inputs: 14 points (X000–X00D)
Outputs: 14 points (Y020–Y02D)

X000 -------------------- X017

T1–40

Y020 -------------------- Y02F

Inputs: 24 points (X000–X017)
Outputs: 16 points (Y020–Y02F)

Any operation for the I/O allocation is not required if only the T1 basic unit is used.

Figure A5.2   Instruction set and data used by Prosec T1 series PLCs (*concluded*).

When X005 is ON or the data of D0100 is greater than 200, Y027 comes ON.
Y027 stays ON even if X005 is OFF and the data of D0100 is 200 or less.
Y027 will come OFF when X006 comes ON.

Figure A5.3   Using the 'greater than' function instruction.

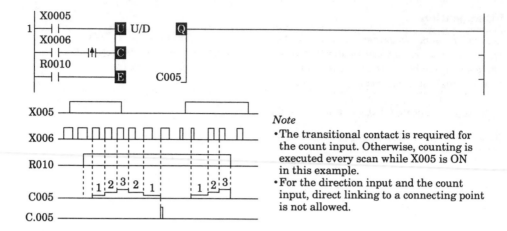

Figure A5.4   Up/down counter.

A5.2.3   DATA TRANSFER

The move (mov) instruction is used for data transfer as shown in Fig. A5.5.

Data transfer

```
  | R0010
1 |——| |——[ 12345 MOV D0100 ]———————————————————————————————————————|
  |
```

When R010 is ON, a constant data (12345) is stored in D0100 and the output is turned ON.

Figure A5.5   Data transfer.

A5.2.4   ADDITION WITH CARRY

An example of adding data stored in double-length registers is shown in Fig. A5.6. The carry flag is set if a carry is generated.

A5.2.5   SUBTRACTION WITH CARRY

An example of subtracting data stored in double-length registers is shown in Fig. A5.7. The carry flag is set if a borrow is generated.

A5.3   Network communications

A TOSLINE-F10 network card can be fitted to the T1-40. This networking capability allows T1-40 units to be linked together to form a small I/O point distribution system as shown in Fig. A5.8 on page 160. The network allows the host computer to read/write PLC data, monitor PLC status and up-load/down-load PLC programs.

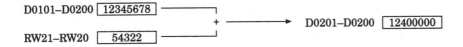

When R013 is ON, the data of double-length registers D0101–D0100 and RW21–RW20 are added, and the result is stored in D0201–D0200. The RSTC is an instruction to reset the carry flag before starting the calculation.

If the data of D0101–D0100 is 12345678 and RW21–RW20 is 54322, the result 12400000 is stored in D0201–D0200.

D0101–D0200 $\boxed{12345678}$ ────┐
                                                    + ────────➤ D0201–D0200 $\boxed{12400000}$
RW21–RW20 $\boxed{54322}$ ────┘

Figure A5.6 Addition with carry.

When R013 is ON, the data of double-length register RW23–RW22 is subtracted from the data of D0201–D0200, and the result is stored in D0211–D0210. The RSTC is an instruction to reset the carry flag before starting the calculation.

If the data of D0201–D0200 is 12345678 and RW23–RW22 is 12340000, the result 5678 is stored in D0211–D0210.

D0201–D0100 $\boxed{12345678}$ ────┐
                                                    − ────────➤ D0211–D0210 $\boxed{5678}$
RW23–RW22 $\boxed{12340000}$ ────┘

Figure A5.7 Subtraction with carry.

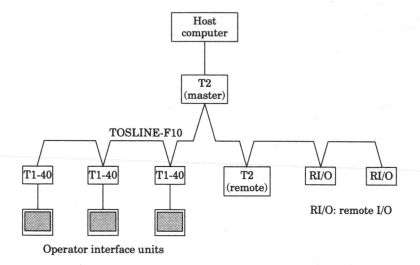

Figure A5.8   Network communications.

Item	TOSLINE-F10 system specifications	
	High-speed mode	Long-distance mode
Topology	Bus (terminated at both ends)	
Transmission distance (without repeater)	500 m max. (total)	1 km max. (total)
Transmission speed	750 kbps	250 kbps
Scan transmission capacity	512 points (32 words) max.	
Scan cycle	7 ms/32 words	12 ms/32 words
Error checking	CRC check	

Toshiba International (Europe) Ltd.
1 Roundwood Avenue,
Stockley Park,
Uxbridge,
Middlesex,
UB11 1AR

Appendix 6
Guide to some leading PLC systems

Company Address	Models	DIO	AIO	Memory (KB)	Scan time (ms/K)	LD/ FB	IL	SFC	Communications
Allen Bradley ♦ Pitfield, Kiln Farm, Milton Keynes, MK11 3DR	Micrologix 100	32		1K	2 ms	Yes			RS232
	SLC 5000 (5/03 CPU)	960	Yes	12K	1 ms	Yes			RS232, RS422 RS485
	PLC-5	3072	Yes	100K	<1 ms	Yes		Yes	RS232, RS422 RS485
GE Fanuc Automation Unit 1, Mill Square, Featherstone Road, Wolverton Mill South, Milton Keynes, MK12 5BZ	Series 90 Micro	14		3K	6 ms	Yes	Yes		RS422, RS485
	Series 90–30	4095	Yes	80K	0.18 ms	Yes	Yes		RS232, RS422 RS485 Ethernet
Klockner Moeller ♦ PO Box 35, Gatehouse Close, Aylesbury, Bucks, HP19 3DH	PS4–100	230	Yes	1K	5 ms	Yes	Yes		Network capability
	PS4-200	1006	Yes	32K	5 ms	Yes	Yes		
Matsushita Automation Controls Sunrise Parkway, Linford Wood East, Milton Keynes, MK14 6LF	FP1	152	Yes	5K	10 ms	Yes	Yes	Yes	Built-in RS232
	FP3/5	2048	Yes	16K	0.5 ms	Yes	Yes	Yes	Networking capability
	FP10	4096	Yes	30K	0.15 ms	Yes	Yes	Yes	SCADA
Mitsubishi Electric ♦ Travellers Lane, Hatfield, Hertfordshire, AL10 8XB	F1	120	Yes	1K	12 ms	Yes	Yes	Yes	RS232 (Medoc)
	F2	240	Yes	2K	7 ms	Yes	Yes	Yes	RS232 (Medoc)
	A1S	512	Yes	8K	1 ms	Yes	Yes	Yes	Networking capability (Melsecnet)
Omron Electronics ♦ 1 Apsley Way, Staples Corner, London, NW2 7HF	CQM1	192	Yes	7.2K words (+6K words)	0.5μs– 1.5μs*	Yes	Yes		RS232, RS485
	CK	148	Yes	2K	10 ms	Yes	Yes		RS232, RS485
	CH(mini-H)	240	Yes	8K	0.75 ms	Yes	Yes		In-built RS232
	C200HS	480	Yes	16K	0.2 ms	Yes	Yes		RS232, RS422, RS485
	C2000H	2048	Yes	32K	0.3 ms	Yes	Yes		SCADA

Company Address	Models	DIO	AIO	Memory (KB)	Scan time (ms/K)	LD/FB	IL	SFC	Communications
Siemens ◆	S5-90U	14		2.2K	10 ms	Yes	Yes	Yes	
Sir William Siemens House, Princess Road, Manchester, M20 2UR	S5-95U	256	Yes	8K	2 ms	Yes	Yes	Yes	Optional modules (e.g. RS232)
	S5-100 (CPU 103)	256	Yes	10K	2 ms	Yes	Yes	Yes	Profibus
	S5-115U	4096	Yes	192K	0.1 ms	Yes	Yes	Yes	
Telemacanique ◆	TSX 07	48	Yes	1K	5 ms	Yes	Yes	Yes	RS485
Sir William Lyons Road, University of Warwick Science Park, Coventry, CV4 7EZ	TSX17	160	Yes	24K	10 ms	Yes	Yes	Yes	RS232
	TSX47-40	1024	Yes	112K	5 ms	Yes	Yes	Yes	Interbus
	SY/MAX 400	4000	Yes	16K	0.7 ms	Yes			Built-in Ethernet interface
Toshiba	M20/M40	20–168	Yes	4K	0.9 μs/contact 110 μs/16-bit addition	Yes			Built-in RS485 Toshline-30 network capability
1 Roundwood Avenue, Stockley Park, Uxbridge, Middlesex, UB11 1AR	EX100	480	Yes	4K	16-bit addition	Yes			
	Prosec T1	200	Yes	2K	2.1 μs/contact 7 μs/16-bit addition	Yes			RS232C and Toshline-F10 (T1–40 only)
	Prosec T2	1024	Yes	9.5K	0.46 μs/contact 2.3 μs/addition	Yes		Yes	RS485 (Host PC) Toshline 30 Toshline F10

The data in this guide has been estimated and compiled from product information. Suppliers should be contacted to confirm latest specifications.

DIO = digital input/output
AIO = analogue input/output
LD/FB = ladder diagram with function block
IL = instruction list (proprietary mnemonics)
SFC = sequential function chart (proprietary implementation)

◆ RS Components Ltd, PO Box 99, Corby, Northants, NN17 9RS (01536 201201) is the sales representative for Allen Bradley, Klockner Moeller, Mitsubishi, Omron, Siemens and Telemecanique.

Index